Writing Math Research Papers
Enrichment for Math Enthusiasts

Robert Gerver

KEY CURRICULUM PRESS
Innovators in Mathematics Education

Writing Math Research Papers
Enrichment for Math Enthusiasts

Author: Robert Gerver, Ph.D.
North Shore High School
Glen Head, New York

Editors: Dan Bennett, Leslie Nielsen

Design/Production/Illustrations: Joe Spooner

Cover: Luis Shein

Publisher: Steven Rasmussen

Editorial Director: John Bergez

Key Curriculum Press
P.O. Box 2304
Berkeley, CA 94702
(510) 548-2304
editorial@keypress.com
http://www.keypress.com

Printed in the United States of America ISBN 1-55953-278-5
10 9 8 7 6 5 4 3 2   00 99 98 97

For my treasures, Linda, Julianne, and Michael

Acknowledgments

Without the help, encouragement, and support of the following people, this book would not have been possible:

William Wagner
Joan Fowler
Dan Bennett

Leslie Nielsen
Beth Porter
North Shore School District

Thanks are given to these North Shore High School students for contributing their fine work, which is featured throughout this book:

Lauren Adelman
Simran Bagga
Dalita Balassanian
Tim Bramfeld
Jeff Chew
Jocelyn Chua
Nancy Friedlander
Blake Gilson

Parisa Golestaneh
Robin Hadley
Laura Henning
Kim Martini
Nicole Miritello
Sharon Poczter
Daniele Rottkamp
Alexis Soterakis

Writing Math Research Papers
Enrichment for Math Enthusiasts

Contents

Introduction

As an avid fan of writing and mathematics, I began giving my students writing assignments as a new teacher in 1977. The initial reaction was "This is *math*—why are we doing English?" As the students completed their math writing projects on varied topics such as automobile insurance, income taxes, stocks and bonds, consumer credit, and more, they began to realize the value of being able to communicate the powerful results mathematics can provide. Students who wrote about surface area, parabolas, the Pythagorean theorem, and other mathematics topics were quick to point out that they internalized the material so much better because in order to explain it, they *really* had to understand it. Students concluded that writing is an essential component of any discipline, not just humanities courses. Based on the successes of these early writing experiences, I began to incorporate writing in my mathematics classes on a daily basis. Needless to say, I was very excited when the National Council of Teachers of Mathematics (NCTM), an organization of educators dedicated to the improvement of mathematics instruction, recommended the integration of writing across the mathematics curriculum. Mathematics and writing, to me, have always complemented each other naturally.

As problem solving became a central focus of math classes, it became clear that writing would need to be integrated with mathematics courses. Problem solvers have to be able to explain the procedures they used and the solutions they found.

Have you ever been absent from a class and asked a fellow student "What did we do in math yesterday?" only to hear the response "I know it but I can't explain it"? How did you react? Imagine a detective telling a superior "I've solved the case but I can't explain my solution"!

Mathematicians must be able to communicate their findings to others so their important results can be used to solve future problems. This is not the only value of writing in mathematics. Writing will also help *you* understand mathematical concepts. As you construct written explanations, you will need to explore and review mathematical concepts in your mind. *Writing Math Research Papers* will help you expand your ability to read about mathematics, explore mathematics, and write about your mathematical experiences.

Why Do Math Research?

If you've been successful in mathematics and you enjoy it, you may wonder: "What other mathematical challenges can I explore? How can I tap my enthusiasm in mathematics to further my education?" Math research will provide you with the opportunity to explore mathematics and enjoy the thrill of discovery. Your exploration of a new concept will give you experience in tackling any nonroutine problem you may encounter. Throughout this book, you will be introduced to the world of research. By learning problem-solving and research skills, you'll vastly increase your potential to solve all types of problems.

You already possess many of the tools necessary to do exploratory mathematics. Perhaps you have done some problem solving in your math class. Your math research paper will be a major project—you will not finish it in one day or even one week. It will require you to read, write, think, and investigate mathematical ideas. It will improve your writing, reading, and oral communication in other subjects as well. It will empower you to create, conjecture, challenge, and question according to your ability, motivation, priorities, and schedule. Research will expose you to the beauty and practicality of mathematics and reward you with the tremendous feeling of uncovering a result previously unknown to you. Communicating your experiences through the writing of the research paper will help you understand and appreciate the mathematics you've explored as well as help others learn about your findings.

Your teacher can serve as coach, and this book can serve as a guide, but the driving force is your own motivation. You will become an expert on your topic because you will spend a great deal of time exploring it. You will learn mathematics by *doing* mathematics.

What's In This Book?

Often, when mathematics "term papers" are assigned to students, students are given a list of topics and a due date. Specific instruction on reading, extending, and writing mathematics may not be offered. Writing mathematics and doing mathematics research involve sophisticated skills that are not innate. The purpose of *Writing Math Research Papers* is to introduce you to these skills and give you direction in developing valuable skills that will last a lifetime. We examine the purpose of each chapter here to better acquaint you with how these skills will be developed.

Chapter 1: Mathematics—Shouting Questions and Whispering Answers

As the title implies, as you engage in mathematical thinking it is natural for many questions to arise. Many of these questions are open-ended, challenging inquisitions. No matter how deeply you delve into a topic, new questions will always surface; hence, mathematics *shouts* questions. Finding the answers to these questions requires time, effort, and skill. The answers do not surface as quickly as the questions; hence, mathematics *whispers* answers. The chapter contains some examples that show that even basic arithmetic concepts can inspire intriguing questions.

Chapter 2: Problem Solving—A Prerequisite for Research

Since the 1980s, problem solving has played an increasingly important role in mathematics courses. You will encounter nonroutine problems on the job and in everyday life, and the solutions to these problems will be crucial. Solid problem-solving experience is advantageous. This chapter reviews problem-solving strategies you may have learned in some of your previous math classes. Familiarity with these strategies will help you answer some of the questions that arise as you carry out your research.

Chapter 3: Writing Mathematics

This chapter provides some tips for writing about mathematics that you can use to improve your writing even if you don't intend to write a math research paper right away. They will improve your note taking in class as well as help you communicate your thoughts clearly whenever you write about mathematics.

Chapter 4: The Math Annotation Project

The Math Annotation Project is a writing activity that will give you practice in writing mathematics before you begin your research paper. It employs the writing tips introduced in Chapter 3 and can be adapted for use in any mathematics class. This writing activity concentrates on mathematics with which you are already familiar so you can concentrate solely on improving your writing skills.

Chapter 5: Conjectures, Theorems, and Proofs

The similarities and differences between history, science, and mathematics research and the role of proofs in each are discussed in this chapter. In your mathematics reading, you will encounter proofs. You should try to read through them and understand them so you can explain them in your paper. You may come up with some questions of your own and even make conjectures (hypotheses) about the answers to these questions. As you become more experienced, you may actually prove your conjectures with original proofs. This chapter will expose you to the role that proofs can play in your research. Your experience with proof will be invaluable to you in future math classes.

Chapter 6: Finding a Topic

Many high school students are first exposed to research in history and science classes. Some students have even written research papers and conducted research experiments in these fields. The difference between a report and a research paper is discussed in this chapter. Often, students who write math papers pick a topic that is too broad. As this chapter points out, topics for research papers must be specific. Mathematics has many sources for succinct topics for high school research papers. Mathematics journals written for teachers and math enthusiasts feature short articles that describe particular mathematical concepts. These articles, usually three to six pages long, can provide an excellent springboard for your paper. Appendix A lists journals and journal articles you can use as sources for your paper's topic. Other suggestions for finding topics are also discussed in this chapter. In summary, your research paper will build upon a short article, problem, or idea, rather than attempt to condense information from several immense library books.

Chapter 7: Reading and Keeping a Research Journal

You might not understand how to turn a short article into a research paper, but as you read through this chapter you will see that you might be able to read fifty pages of a novel in less time than it takes to carefully read and digest one page or even one paragraph of mathematics. Specific tips for reading mathematics journal articles are given. Because you will not start writing the formal paper until much of your research is completed, you will need a comprehensive set of notes on your research. Thus, you will keep a journal as you read.

Chapter 8: Components of Your Research Paper

Chapters 3 and 4 introduce you to writing mathematics, and Chapters 5, 6, and 7 instruct you in how to conduct your research in a logical fashion. Chapter 8 helps you pull it all together for the formal paper. The parts of the research paper are discussed. As you read this chapter, you will also refer to Appendix B, which features samples of actual student work that will help you get a feel for how the formal paper should appear.

Chapter 9: Oral Presentations

Many students are given a chance to present their research at special events and mathematics competitions. Giving an oral presentation requires extensive planning, visual aids, and practice. This chapter will give you tips in staging a polished, professional-quality presentation that reflects the high quality of your research.

After reading the purpose of each chapter, you may wonder: "Will doing a math research paper strengthen my mathematics education? If so, how?" The skills you will acquire during the project are considered very valuable by mathematics educators and can be applied to other mathematics courses as well as to other disciplines.

Math Research: Meeting The Recommendations Of The NCTM

The open-endedness of math research makes it adaptable to the continual changes in mathematics and math education. Have you noticed any such changes in your schools? How much is technology (calculators and computers) a part of your math classes? Do you think your current math textbook is different from one in use twenty years ago? The content—the math that is taught—is often revised. In addition, teaching methods—how that content is taught—are adjusted to meet the current needs of students and society. The National Council of Teachers of Mathematics (NCTM) made many recommendations for the 1990s and beyond in their publication *Curriculum and Evaluation Standards for School Mathematics*. Many of the changes in math education in the 1990s reflect these recommendations. The NCTM proposed that the following four main unifying threads pervade the entire math curriculum (NCTM, 1989, p.123):

1. mathematics as communication

2. mathematics as reasoning
3. mathematical connections
4. mathematics as problem solving.

Your research project reflects an effective, practical incorporation of many of these suggestions in a format that will allow you to experience some of the power of the varied teaching and learning strategies used in the math class. Your research topic may be a mathematical application to another discipline or a "pure" mathematics topic—one that advances knowledge about a certain mathematics topic and whose aim is not to solve a practical problem in another discipline. In either case, the research strategies you learn will be valuable in virtually all mathematical situations you encounter. Chapter 6: Finding a Topic will help you decide what road you should take. You will be traveling down a mathematical highway illuminated by communication, reasoning, connections, and problem solving.

Communication and Your Research

Researchers need to be communications-minded. They must read, attend lectures, listen, write clearly, and make oral presentations of their work. Researchers need to digest information, process it, perform their work, and report their findings. You will benefit from others' ability to communicate because, early in your research, you will be reading about your topic. Chapter 7: Reading and Keeping a Research Journal examines specific reading skills you can adopt to help in reading technical writing. As you read, you will ask yourself informal questions to help you understand passages as well as formal questions as mathematical extensions of your readings. Your questions must be clear to outside readers. Good communication skills are essential to effective presentations, both written and oral. Chapter 3: Writing Mathematics, Chapter 4: The Math Annotation Project, Chapter 8: Components of Your Research Paper, and Chapter 9: Oral Presentations delve into the communication arts with respect to your research. Above all, since clarity is paramount, your organization, definitions, questions, proofs, conjectures, and explanations must be well-written and logically organized. You'll need to reflect on each stage of your work carefully before you can put it into your own words. You'll need to use notation effectively in order to convey mathematics succinctly. Becoming adept at communicating your work will allow others to benefit from the knowledge you've acquired. Other student-researchers may want to continue your work, extend it, alter it, or make new investigations. They will rely on your communication skills.

Reasoning and Your Research

As you read through your research articles, you will need to follow the arguments presented by the authors. You must make sure that their arguments are valid and that you understand the logic used by testing their claims and working through their proofs step-by-step. You will formulate ideas based on patterns, your mathematical intuition, and the mathematical tools you have acquired in school. Making conjectures requires reasoning—conjectures aren't guesses but rather hypotheses that, whether true or false, can reasonably be tested. Mathematicians make many conjectures that turn out to be false. If, after working with a conjecture, you suspect that it is false, you might try to find a counterexample—a single case in which the conjecture is not true— or explain theoretically why your original suspicions were not true in every case. If you are convinced that your conjecture is true, you may try to construct a proof—a valid argument that your hypothesis is indeed a theorem. There are different types of proofs for different conjectures. You might read a statement that is not proved and decide to construct a proof on your own. This proof then becomes part of your research. Chapter 5: Conjectures, Theorems, and Proofs is an overview of the use of proofs in mathematics.

Connections and Your Research

As you explore your paper's topic, you will have the chance to integrate different branches of mathematics in your research. Students doing research on a geometry topic might use algebra, logic, calculus, trigonometry, set theory, and more to create proofs and give explanations about their topic. Tapping the different fields as they are needed requires knowledge and discretion, as well as a well-equipped mathematical tool kit. In this respect, your paper is different from the study of a single unit in math class. You may need to learn part of a topic on your own (with the assistance of your teacher and an appropriate textbook) because it can help your research. You may be able to test or prove one of your conjectures in two different ways; for example, using coordinate geometry or plane geometry. As part of your research, you might discuss which method was easier, better, faster, shorter, more intuitive, and so on.

   If your paper deals with a mathematics application to another discipline, you will be making connections not only within mathematics but between mathematics and the discipline you're modeling. When researching an applied-math topic, you need to become knowledgeable about the discipline you are researching as well as the mathematics you are using. The connections are seemingly endless:

- ❏ What must an architect know about an ellipse in order to design a whispering gallery?
- ❏ What is the shape of the suspended cables of a suspension bridge?
- ❏ How are seismographs, rates, and circles used to find the epicenter of an earthquake?
- ❏ How can mathematics be used to find the area of an irregular shape such as a golf green?

❏ How are graphs and statistics used to predict economic trends?

❏ How can doctors use conic sections to break up kidney stones without invasive surgery?

As society becomes more technologically oriented, mathematics assumes a more prevalent role in the progress of other disciplines. A connection to mathematics is an essential component of the research that will advance knowledge in other fields. Use your communication skills to help your readers make connections *in* your work and *to* your work.

Problem Solving and Your Research

Your readings will include passages that you don't immediately understand. As you re-read certain sentences several times, you will need to employ your problem-solving skills to figure out their meaning. The passage you are having trouble with in effect becomes its own problem. Attack it with determination. Your readings will include many such hurdles. You might even create some hurdles yourself, since each concept you understand and internalize may breed more questions and possible extensions. Such questions and extensions are really new problems. Pose them to your readers to investigate on their own, or raise their solutions and address their solutions in your research. Chapter 2: Problem Solving—A Prerequisite for Research provides you with an overview of problem-solving strategies.

How To Use *Writing Math Research Papers*

Ideally, you will use *Writing Math Research Papers* in a group setting, with your classmates and teacher giving you feedback and coaching. If you are planning to write a math research paper, you and your classmates should read *Writing Math Research Papers* gradually as you go through the research process. As you progress through each stage of your research, you should reread previous chapters and use them as a reference. The following time line is offered as a general guide to your project. It is based on a nine-month school year, but can be adjusted to meet your specific situation. The time line can help you prorate your time if you are not writing your paper over a full school-year period. The amount of time you have, the amount of coaching you receive, and the topic you choose will affect the amount of time you spend on any one activity.

Month	Activities
1	Read and discuss *Writing Math Research Papers,* Chapters 1 through 6. Find a topic and an article. Make copies of your article, and start a Math Annotation Project.
2, 3, 4	Read and discuss Chapter 7. Finish your Math Annotation Project. Complete the Journal Article Reading Assignment. Begin reading and annotating your article. Begin your research journal. Begin consultations with your teacher.
5, 6, 7	Continue research and consultations. Read Chapter 8 and review Chapters 3 and 4. Begin making an outline for your formal paper. Start writing, submitting, and revising drafts of the formal paper.
8	Stop any new research and spend time polishing the formal research paper. Submit drafts, make corrections, and discuss the formal paper at your teacher consultations.
9	Read Chapter 9. Plan and make an oral presentation.

Following are some suggestions for other ways you could use the book:
❏ If you are not sure whether or not you are going to write a research paper, reading through Chapters 1–6 can help you decide if you should undertake such a project.
❏ If you want to improve your mathematics writing skills by practicing writing about the mathematics you have learned in school, read Chapters 3 and 4.
❏ If you would like to improve your ability to take notes in mathematics classes, read and do the activities in Chapter 4.
❏ If you are a teacher or administrator planning a course in problem solving and/or mathematics research, read the entire book, with an emphasis on Appendix C: A Guide for Teachers and Administrators.
❏ If you are a teacher coaching a student who is doing an independent-study math research paper, Chapter 6 and Appendix A will help you and your student find a topic. Appendix C will help you understand the role you can play as this student's mentor.

As you enter the world of mathematics writing and mathematics research, keep in mind that fellow teachers and students are always interested in your ideas, successes, and suggestions. Your comments can be sent to: Robert Gerver, c/o Key Curriculum Press, P.O. Box 2304, Berkeley, CA 94702

Write on!

Chapter 1

Mathematics—Shouting Questions and Whispering Answers

Examine the following set of events:

$$tan\left(\frac{\pi}{4}\right) = \frac{\frac{1}{\sqrt{2}}}{\frac{1}{\sqrt{2}}} = 1$$

❑ Which element seems to stand out?
❑ Which element interests you the most?
❑ Can you think of a way in which each element relates to the other three?
❑ Can you think of something these four elements have in common?

Although the answers to these questions may vary, you might recognize several traits that all four elements have in common:

❑ They engage millions of people.
❑ They are challenging for the participants.
❑ They can be exciting.
❑ Each has a degree of uncertainty.
❑ Each offers an opportunity for teamwork.
❑ There is exhilaration and satisfaction when a given task in each element is completed successfully.
❑ There are many unknowns in each scenario.

Perhaps you think of algebra when you hear the term "unknowns." Often in algebra, we are trying to find the unknown value of a variable. In our current context, we are using the term "unknowns" to talk about problems, facts, and theories that are currently unsolved or not yet discovered. Why are we talking about unknowns in mathematics? Hasn't all mathematics already been discovered? The answer to this question is a resounding no! There is much yet to be uncovered in all branches of mathematics. Note that the experience of exploring the unknown does not have to deal with unsolved problems. A problem that is new to *you* is an unsolved problem for you. Much beautiful, fascinating, surprising, and useful mathematics is new to you and waiting to be explored.

What Mathematics Is Unknown to *You?*

Most likely, your research, especially at the beginning stages, will be confined to mathematics that is known in the field but not known to you specifically. The journal articles referred to in the Introduction are a terrific source for such mathematics topics. Let's examine a problem about consecutive integers and their sums from a journal article (Olson, 1991). Notice that the following numbers can be expressed as sums of consecutive integers:

$$21 = 6 + 7 + 8$$
$$5 = 2 + 3$$
$$30 = 4 + 5 + 6 + 7 + 8$$
$$2 = (-1) + 0 + 1 + 2$$
$$11 = (-10) + (-9) + (-8) + (-7) + (-6) + (-5) + (-4) + (-3) + (-2) + (-1) + 0 + 1 + 2 +$$
$$3 + 4 + 5 + 6 + 7 + 8 + 9 + 10 + 11$$

Can *any* whole number be expressed as the sum of consecutive integers? Can any whole number be expressed as the sum of consecutive counting numbers? What do you think? Try expressing the counting numbers from 1 through 10 as sums of consecutive integers. Can you express any of these numbers as consecutive-integer sums in more than one way? Although the answers to these questions are "known," if the answers are unknown to you, you can investigate these problems as unsolved problems in mathematics. You will truly be "doing" mathematics as you try to uncover the solutions. Let's continue by investigating properties of a very famous sequence.

Perhaps you are familiar with the Fibonacci sequence. The Fibonacci sequence is a sequence of positive integers that begins with 1, 1. Subsequent terms are found by adding the two previous terms. The first fifteen terms of the Fibonacci sequence are listed here:

$$1, 1, 2, 3, 5, 8, 13, 21, 34, 55, 89, 144, 233, 377, 610 \ldots$$

What patterns do you notice? Can you create questions about the sequence? Let's look at some patterns and questions to become accustomed to viewing mathematics as a very "open" rather than a "closed" science.

Some questions:
❏ Are there infinitely many Fibonacci numbers?
❏ Are there more odd than even Fibonacci numbers?
❏ Are any Fibonacci numbers perfect squares?
❏ Are any Fibonacci numbers prime?

Some patterns:
❏ The numbers are increasing.
❏ Every fifth Fibonacci number is divisible by 5.
❏ The difference between consecutive Fibonacci numbers increases as the numbers increase.
❏ The sequence seems to have two odd integers followed by one even integer.
❏ If two consecutive Fibonacci numbers are squared, the sum of these two squares is also a Fibonacci number.
❏ The ratio of one term to the previous term is always less than or equal to 2.

How many of the patterns did you notice? What other patterns did you notice? How could conjectures based on these patterns be proved? Do you think there are other patterns you haven't noticed yet? The answers to questions, the formulation of conjectures, and the proofs of those conjectures may not be uncharted territory in mathematics, but they may be unsolved problems for you. Solving problems that are original for you allows you to experience the discovery process. The freshness of new, often fascinating ideas can team up with the mathematics you have already learned in school to provide you with a challenging, rewarding trip into the world of real mathematics!

Becoming Inquisitive

As you ponder virtually any mathematics topic, questions will naturally arise. Even seemingly simple ideas contain unlocked secrets. The purpose of this section is to show you that questions exist everywhere and that you should always formulate questions about the mathematics you are doing. The main issue here is the questions; in fact, some of the answers are deliberately omitted. You might try to answer them as part of a research project or simply as a problem-solving exercise. Concentrate on the *questions* in each example. Try to think of your own questions, too. The first example involves elementary subtraction.

Take a look at basic subtraction of three-digit integers by examining the following subtraction example:

$$
\begin{array}{r}
954 \\
-459 \\
\hline
495
\end{array}
$$

Did you notice that the minuend, the subtrahend, and the difference were all formed from the same three digits, 4, 5, and 9? For what other three-digit numbers does this happen? Did you notice that the minuend's digits were in descending order and the subtrahend's in ascending order? How many three-digit numbers are there? Must all of them be tested individually to answer the first question? Do the outermost digits of the difference always add up to the middle digit? Did you ever think that a simple subtraction example could launch so many questions? Can you think of other questions based on this example?

When you were in elementary school, you learned how to multiply fractions:

$$\frac{1}{2} \times \frac{3}{5} = \frac{3}{10}$$

The procedure was so predictable that many students were able to guess it without prodding from the teacher. After learning multiplication, you learned how to divide fractions:

$$\frac{25}{32} \div \frac{5}{8} =$$

Based on the multiplication algorithm, your first guess might be to divide the numerators and then divide the denominators. Is this correct?

$$\frac{25 \div 5}{32 \div 8} = \frac{5}{4}$$

The traditional procedure requires you to find the reciprocal of the divisor and multiply:

$$\frac{{}^{4}\cancel{25}}{{}_{5}\cancel{32}} \times \frac{\cancel{8}^{1}}{\cancel{5}_{1}} = \frac{5}{4}$$

Notice that the answers match. Is this a coincidence, or is the first procedure valid? Does it always work? The first procedure was correct! Then why is the more unnatural, second procedure a staple in elementary schools? Does the second procedure have advantages? In fact, the first procedure always works, but it can become cumbersome when the numerator and denominator of the divisor are not factors of the numerator and denominator of the dividend, respectively, as shown in the following example:

$$\frac{12}{25} \div \frac{9}{10} = \frac{12 \div 9}{25 \div 10} = ?$$

$$\frac{{}^{4}\cancel{12}}{{}_{4}\cancel{25}} \times \frac{\cancel{10}^{2}}{\cancel{9}_{3}} = \frac{8}{15}$$

There is a logical reason for the universality of the second procedure; however, there is plenty of room for conjectures, trials, errors, questions, and interesting discussion. It is never too early to develop a feel for questioning and probing. Today's mathematics classes are encouraging this.

The procedure for addition of two fractions involves common denominators. Most youngsters' best guesses of the sum of two fractions probably resemble the example below:

$$\frac{4}{5} + \frac{2}{3} = \frac{6}{8}$$

We know this is not the correct answer. Why not? Would this method ever work? For what fractions? Can you think of a scenario in which the above arithmetic algorithm might make sense?

> Julianne received 4 out of 5 on yesterday's French quiz. She received 2 out of 3 on today's French quiz. What fraction can be used to convert her cumulative quiz scores to a percent for the two quizzes?

The answer is 6/8. On the number line, we know that 4/5 + 2/3 ≠ 6/8. However, the intuitive but incorrect attempt of this addition problem does have an application. Perhaps a symbol other than + could be created to indicate when the above algorithm can appropriately be applied to two fractions.

Learning that something *isn't* true is a valuable part of probing in mathematics. It creates a natural motivation to search for a correct procedure. If after much trial, error, and questioning you discover a correct procedure on your own, you experience *real* mathematics. Let's examine elementary fractions further.

Remember when you learned how to "reduce" fractions? The word is in quotes because we don't make a fraction smaller when we reduce it but rather make the numerator and denominator smaller. This procedure is often referred to as "simplifying" the fraction. We say that 11/33 is equal to 1/3 in simplest form. Even this term is sometimes misleading. You could argue that, although 19/25 is the "simplified" version of 76/100, in fact 76/100 is "simpler" to understand because we are so used to comparing numbers to 100 (converting to percents). If you answered 19 out of 25 questions correctly on a quiz, the first thing you would probably do is convert it to 76% to get a better feel for how you did. Would you now question which fraction is actually simpler? Did you question this in elementary school? Look at the following procedure for simplifying fractions:

$$\frac{1\cancel{6}}{\cancel{6}4} = \frac{1}{4} \quad \text{and} \quad \frac{1\cancel{9}}{\cancel{9}5} = \frac{1}{5}$$

Are the answers correct? Yes. Does this procedure always work? When does it? How would you attempt to answer these questions?

Let's examine a problem involving decimals and percents.

> Carl bought a lawn mower that was discounted 25% because it was a damaged floor model. The store was also having a 10%-off sale. The cashier first took 10% off the original price, and then took 25% off the reduced price. Carl requested that the store first take 25% off the original price and then 10% off the reduced price, since he felt that 25% of the larger price would bring a greater discount.

Does it matter which discount is taken first? Is Carl's discounted price equivalent to a 35% discount? How would you attempt to answer these questions?

Doing math research requires you to become adept at asking questions. Math research is a series of questions, conjectures, and answers based on the investigation of a topic. The chance to make conjectures is a building block of good problem solving. Did you realize that the mathematical discoveries with which you are familiar took many attempts to discover? Can you imagine the elation and satisfaction people feel when they encounter a problem, create key questions, answer the questions, and solve the problem? A great deal of challenge and excitement awaits you as you enter your mathematical unknown via your research project.

Unknowns in Mathematics

Although your research will not focus initially on solving unknown problems in mathematics, you might be surprised and excited to find out that many unsolved problems can be explained on a very elementary level. There are many unsolved problems in mathematics. Perhaps you think that unsolved problems exist only in the very high levels of advanced college mathematics. That is a misconception. All areas of mathematics have unsolved problems. We will investigate three unsolved problems related to prime numbers.

Prime numbers are numbers that have exactly two divisors—themselves and the number 1: {2, 3, 5, 7, 11, 13, 17, 19, 23, 29, 31, 37, 41, 43, 47, 53, 59, 61, . . .}.

Are there infinitely many prime numbers? What do you think? How would you go about answering this question? Would your answer be influenced by a person giving you a computer-generated list of the first 25,000 prime numbers?

There are, in fact, infinitely many prime numbers. The proof is a standard part of a number theory course and is not typically encountered in the high school curriculum. (It can, however, be understood by a high school student and could be one part of a research paper on number theory.) Take another look at the list of prime numbers above. You could make a list of things you notice. For example:

❑ There is only one even prime number.
❑ No primes are perfect squares.
❑ If a prime is equal to the sum of two other primes, one of the addends must be 2.
❑ The distance between consecutive primes does not strictly increase as the primes get larger.

There are many more patterns to be discovered. Look at the following pairs of primes, called twin-prime pairs:

$$3, 5$$
$$5, 7$$
$$11, 13$$
$$17, 19$$
$$29, 31$$
$$41, 43$$
$$59, 61$$

Notice that we are able to list consecutive odd numbers that are prime to form twin-prime pairs. Can you add to this list? What patterns do you notice in the list of twin-prime pairs? Do you think there are infinitely many twin-prime pairs? The answer to this last question is unknown—it is an unsolved problem in mathematics! As you can see, unsolved problems do not necessarily have to be hard to understand, even though solving them may be challenging. Be aware that a list of thousands of twin-prime pairs, possibly generated on a computer, does not guarantee the existence of infinitely many twin-prime pairs. Such a list would certainly be helpful in making conjectures and looking for patterns, but it would not constitute a proof that there are infinitely many twin-prime pairs.

Do you think there is a formula that will generate only prime numbers? Examine the expression

$$y = x^2 - x + 41$$

Table 1.1 lists x-values from 0 to 41 and their corresponding y-values. (See next page.)

You may have noticed that all of the numbers generated are odd. Will this always be true? If we factor part of the expression, we find that

$$x^2 - x + 41 = x(x - 1) + 41$$

Table 1.1 Ordered Pairs Generated by the Formula $y = x^2 - x + 41$

x	y	x	y	x	y
0	41	14	223	28	797
1	41	15	251	29	853
2	43	16	281	30	911
3	47	17	313	31	971
4	53	18	347	32	1033
5	61	19	383	33	1097
6	71	20	421	34	1163
7	83	21	461	35	1231
8	97	22	503	36	1301
9	113	23	547	37	1373
10	131	24	593	38	1447
11	151	25	641	39	1523
12	173	26	691	40	1601
13	197	27	743	41	1681

The product $x(x - 1)$ is always even, since x and $x - 1$ are consecutive integers. Do you know why? When 41 is added to this product, the sum value will be odd. The expression $x^2 - x + 41$ is certainly an odd-number generator. Look at the first forty-one y-values generated by the formula ($x = 0$ to 40). These values are all prime. Based on the fact that the formula generated a prime the first forty-one times it was applied, would you believe that the formula is a prime number generator? Do you "trust" patterns that work for many examples and believe that they will *always* work? When $x = 41$ in this formula, $y = 1681$, a composite number. Therefore, the formula is not a prime number generator.

This result could have been found by trial and error with a calculator or by inspecting the formula and noticing that when 41 is substituted for x all three addends are multiples of 41 and as a result, the sum is divisible by 41 and is not prime. You may have noticed this from the start! Mathematicians have not yet found a prime-number generator—this is another unsolved problem in mathematics.

Let's investigate our third unsolved problem regarding prime numbers. Examine all of the even integers greater than or equal to 4. See if you can express them as the sum of two prime numbers:

$$4 = 2 + 2$$
$$6 = 3 + 3$$
$$8 = 3 + 5$$
$$10 = 3 + 7$$
$$12 = 5 + 7$$
$$14 = 3 + 11$$
$$16 = 5 + 11$$

The statement "Every even integer greater than or equal to 4 can be expressed as the sum of two prime numbers" is known as Goldbach's Conjecture, and it is not known whether the statement is true or false. Can you tell why the theorem does not hold for all *odd* integers? (Hint: The only even prime number is 2.)

Questions and Answers—Beautiful and Practical

Mathematics is an extremely useful science, with applications in health, economics, astronomy, physics, sports, medicine, and so on. However, like art and music, mathematics is beautiful in its own right—pure mathematics can be just as inspiring as the Mona Lisa. Can you see now that mathematics really does have much in common with baseball, rocket launches, and rock concerts?

> "The true spirit of delight, the exaltations, the sense of being more than man, which is the touchstone of excellence, is to be found in mathematics as surely as in poetry."
> —Bertrand Russell, *A History of Western Philosophy*, 1945

Throughout this chapter, you "heard" mathematics shout questions. The many questions offered here are just the tip of the iceberg. Because mathematics only whispers answers, Chapter 2 focuses on problem solving, which will help you "hear" answers. The thrill of experimentation and discovery awaits!

Chapter Two
Problem Solving—A Prerequisite for Research

> Solving problems is a practical art, like swimming, or skiing, or playing
> the piano: you can learn it only by imitation and practice. . . . if you wish
> to learn swimming you have to go into the water, and if you wish to be-
> come a problem solver you have to solve problems.
>
> —George Polya, *Mathematical Discovery*, 1961

As the world gets politically, economically, and technologically more complex, problem solving has become a primary focus of schools, government, and businesses. The rote skills taught to students in years past would not have equipped them to solve the problems facing the world today. Think of some differences between today's world and the world of fifty years ago, and discuss answers to the following questions with your classmates before reading further.

❑ What is problem solving?

❑ Why is problem solving a sensible focus for schools?

❑ What role do you think problem solving plays in math research?

❑ Why do businesses and the government need problem solvers?

❑ Why is it so difficult to solve global problems?

Many of the problems facing the world today are new, nonroutine problems. How will you handle the problems you encounter? Is there a *routine* for researching the solutions to *nonroutine* questions? You won't be able to simply mimic old solutions. You will face problems that haven't even been conceived of yet, so the problem-solving strategies you learn must be extendible. The strategies and tactics you study cannot consist of mere facts—they must be *procedures* that can be used in all or most problem-solving situations. As a discipline, problem solving is a collection of strategies that can be used to find a solution to a specific situation called the **problem**. These problem-solving strategies apply to all situations, not just mathematics.

This chapter serves as an abbreviated refresher for people who have had problem-solving experience and as an overture for people new to the problem-solving arena. If you are an experienced problem solver, you know that the accumulation of strategies, tips, and pitfalls will strengthen your skills. This chapter will benefit you in that sense as well. An excellent, thorough introductory treatment of problem solving can be found in *Problem Solving Strategies—Crossing the River With Dogs*, by Ted Herr and Ken Johnson. George Polya's *How to Solve It* is also a must-read for the serious problem solver. These and other books on problem solving are listed in Appendix A. Look for some of these books in your library; they will help you sharpen your problem-solving skills.

The skills you acquire in a problem-solving course will have a direct carryover to your mathematics research. You may have thought of a mathematics research paper as a "math book report." As you read through this book, however, you will deepen your understanding of the difference between writing a report and doing research. You will have a firsthand look at the interdependence of problem solving and math research. Your research will lead you to many nonroutine situations; persistence and problem-solving skills will optimize your chances of arriving at a solution. Throughout your research, you will encounter claims and statements that you don't understand. You will discover patterns and formulate conjectures that you want to prove or find counterexamples for. You will often be in the position of trying to decipher and master something you have never seen before. The problems that form the cornerstone of a problem-solving curriculum are small capsules of the hurdles you'll meet as your research progresses. The strategies used in problem solving are effective strategies for attacking these obstacles, because they chart a systematic path through uncharted territory.

Problem solving is inherent to all branches of mathematics, including geometry, probability, algebra, and calculus. The research and problem-solving skills you acquire through mathematics will help you solve not only purely mathematical and math-related problems but also nonmath problems, because the essence of problem solving is a logical, systematic process that leads to a solution. An analogy can be made to the changing of a flat tire. A road service can be called to the scene of a flat tire, and a mechanic can change the tire for the driver who does not know how to do it. The driver's problem is solved. The very same mechanic can teach the driver about the tools and safety procedures involved in changing a flat tire. Equipped with these extendible skills, the driver can safely change flats in future roadside emergencies without assistance and on different cars. In addition to having the problem solved, the driver has learned *how* to solve the problem. Learning strategies is more effective than learning answers, because strategies can be extended while an answer is applicable only to one specific question.

Our Quantitatively Oriented World

Take a look around at the technological world you live in. Look at the new discoveries made in your lifetime alone. Look at how the workplace is changing. The demands of the twenty-first century will require a citizenry well versed in problem solving and in meeting nonroutine, unanticipated demands both at home and in the workplace. Imagine these scenarios:

> A town is planning to construct a domed stadium. You are in charge of the construction of the dome. You have many factors to consider. How much

weight must the dome support? How much do snow and rain weigh? How will wind affect the dome? How will expansion and contraction due to temperature changes affect the dome? What are the least expensive materials that will meet your demands for strength? How does the probability of more severe, but rare, stress and weight circumstances affect your decision as to whether to build a stronger, more costly dome? In answering these questions, you will face a constant battle of cost versus quality of materials and workmanship. Would your dome stay structurally sound after a 10-foot snowfall? What is the probability of such a storm in your area? The solutions to these science, economics, and safety issues all incorporate mathematics.

You are a consultant to a team of scientists that is on the verge of finding a cure for a major disease. The cure involves securing materials from tropical rain forests. Environmental, political, health, cultural, and sociological concerns are involved. The debate is laced with mathematics. How much of the medicine is needed? How much of the rain forest needs to be depleted? What will be the environmental impact of this depletion? How long would it take to renew the depleted acreage? How will the medicine affect the life span of ill individuals? What impact will the increased life span of afflicted individuals have on available medical facilities?

Your company has created an automobile engine that lasts four times longer than the current breed of engines without needing any repairs. How much more will the engine cost the consumer? How drastically will it affect the auto repair business? What changes need to be made to the rest of the car so it can last long enough to take advantage of extended engine life? The domino effect on automobile manufacturers and related industries has a major impact on the economy of the entire nation.

Try to think of your own scenarios that require nonroutine problem solving in our increasingly quantitative-oriented society.

The Routine vs. the Nonroutine

Have you ever tried to do a homework problem, encountered difficulty, and said to yourself, "We never did one like that in class!" Maybe you thought it was unfair to be given such an assignment. Schools provide you with problem-solving training by exposing you to many problem situations in your classes. You must learn how to extend and apply, with discretion, the tremendous amount of knowledge you have amassed during your school years.

Let's use an example to show the difference between doing a typical homework example and solving a problem.

When you learn how to factor trinomials, you might drill yourself by doing five, ten, or 100 examples and, through this practice, become adept in this one skill. We wouldn't usually consider the 101st trinomial factorization you try, say, $x^2 - 4x - 5$, a "problem." The first one you try, if you factor it on your own, without benefit of the teacher's lesson, might be considered a problem. Though small in scope, it may represent a new, nonroutine situation for you for which you employ some problem-solving strategies and content knowledge to arrive at an answer. *Could* the 101st trinomial factorization be considered a problem? Yes, if it involves a new situation. If the coefficients make finding a correct factorization less routine than it was in the previous examples, it is a problem. Look at the example below:

$$\text{Factor: } 36x^2 - 255x + 100.$$

The large number of combinations of factors of 36 and 100 may require you to do some ground breaking above and beyond the skills required to factor $x^2 - 8x + 15$. In that sense, this factoring question represents a nonroutine situation. (Try to factor it!)

Problem solving classically involves nonroutine situations, such as the following example, whose solution depends on the content knowledge of factoring.

Problem: Prove that if x is a positive integer, $5x^3 - 5x$ is always divisible by 30.

Solution: If $5x^3 - 5x$ is divisible by 30, then $5(x^3 - x)$ is divisible by 30, and $x^3 - x$ must be divisible by 6. The expression $x^3 - x$ can be factored as follows:

$$x^3 - x = x(x^2 - 1) = x(x - 1)(x + 1)$$

These three factors can be commuted:

$$x(x - 1)(x + 1) = (x - 1)x(x + 1).$$

Notice that if x is an integer, the product above is the product of three consecutive integers. The product of any three consecutive integers is always divisible by 3, because one of the integers must be a multiple of 3. Also, at least one of the integers must be even, that is, divisible by 2. As a result, $x^3 - x$ is divisible by 6. We can now state that the difference between a positive number and its cube is always divisible by 6. Consequently, $5x^3 - 5x$ is divisible by 30.

Would you consider the problem and its solution routine? Would you have solved the problem in the same amount of time it took you to factor a dozen trinomials? A contradiction arises. How can the nonroutine become routine? How can anybody, ever, become a good problem solver if, in fact, the next problem he or she meets does not mimic one already seen? The answer to this question lies in the quote by Polya above and in the fact that, when you learn how to problem-solve, you are learning *approaches* to solving problems. "Problem solving has been defined as what to do when you don't know what to do" (Herr & Johnson, 1994, p. 2). If you learn a number of approaches and practice them on enough problems, you'll learn how to apply these approaches to new situations.

A General Approach to Problem Solving

George Polya's classic book on problem solving, *How to Solve It*, was first published in 1945. One-half century later, the timelessness of the book remains a credit to Polya's practical approach to solving problems. You should find this book in your library and take the time to read it. In the book, Polya describes four steps critical to the solution of any problem:

1. Understand the problem.
2. Devise a plan to solve the problem.
3. Carry out the plan.
4. Look back and check the result.

A well-organized problem solving strategy is a superb research tool. We will use the acronym SUPERB to embellish and explain Polya's problem-solving steps and make them easy to remember.

S: **Scrutinize** the problem. Read the entire problem thoroughly to the end. Don't skim. Don't make assumptions. Be open-minded. Begin by mentally sorting out the given information. Note if too much information is given or not enough.

U: **Underline** key phrases, questions, requirements, conditions, and so on. In problem solving and math research, it is always wise to have a colored highlighter on hand and to use photocopies rather than original sources so you can write on the problem. Draw key diagrams. Express key phrases in math notation.

P: **Plan** an appropriate problem-solving strategy. You may know some already. Several problem-solving strategies will be discussed later, and you may plan to use one strategy or a combination. Don't worry about specific data; simply formulate a general outline of what you will do to solve the problem.

E: **Execute** your plan. Carry out the procedure you set up for your solution and find the solution. Save all of your work, even the dead-ends. Try to be organized, so that if you stop work in the middle of a process, you'll understand your previous work when you return to it. It will also be easier to check the accuracy of each step you took if your work is organized. Often, problem-solving attempts are done in a scratch-work style, and that is fine, but make sure you can decipher what you have done! If at some point the work must be made comprehensible to an outside reader, your ability to write and explain the solution becomes critical. The solution must be written in complete sentences.

R: **Review** the work you've done. Check to see if your answer makes sense. Make sure you answered the question in full sentences. Review your steps and any arithmetic or algorithms for correctness. Check that you applied the correct data. Determine whether your solution contradicts any of the conditions of the problem.

B: **Build** this solution into your problem-solving arsenal by writing out all the steps of the solution in full sentences. Write out the problem-solving strategies you used. This will help you sharpen your writing skills for your research and give you practice in expressing your findings so they can be understood by others. It will also help you recall how you solved the problem if you return to it in the future to help solve a related problem. List a few questions that come to mind about the problem, or create a new question based on the problem. Write clearly! Something that seems obvious to you as you discuss the problem with classmates now may not be so obvious weeks from now.

Why Scrutinize?

The scrutinize step augments Polya's "understand the problem" step. It is designed as a precaution. Too often, good mathematics students attack their schoolwork, especially timed exercises, with a voracious appetite. While admirable, this often results in starting problems before reading them carefully, anticipating the question that will be asked before it is asked, and attempting a solution without relaxing, focusing, and re-reading the problem. Try not to be too anxious when you approach a problem. Reading a problem is different from reading a magazine article. Also, not all problems come from books. Some come from real-life scenarios, even your own experiences, and might include extraneous information, not enough information, or incorrect information. The scrutinize step is a reminder to focus on relevant information in each problem. Scrutinize and then solve the following problem.

Jacquie's monthly income of $1460 is allocated as shown in the following circle graph:

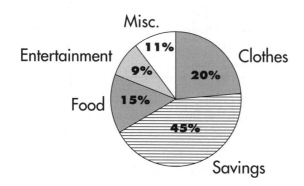

How much of Jacquie's income goes toward her greatest expense?

A superficial look at the problem would lead you to take 45% of $1460 and report that the answer is $657. A closer look might cause you to ask the question, Do we consider savings to be an *expense*? If not, clothes are her greatest expense, and the answer is 20% of $1460, or $292. This issue should be resolved before the mathematics is performed. You might not have noticed the ambiguity without careful scrutiny.

Let's take a look at how crucial scrutinizing the following problem carefully is.

> Eve is planning a rectangular flower garden that will contain a 36-square-foot flower bed. She wants to enclose the garden with a short decorative fence that is sold in 1-foot lengths that snap together. Each 1-foot length costs $3.99, plus 8% sales tax. Find the dimensions of the garden that requires the least amount of fence.

This problem will be solved later in the chapter. Note that the solution is not the cost of the fence, although that could be inferred in a quick read of the problem. Such inferences lead to wasted time and effort and incorrect answers. Careful scrutiny helps you avoid such errors.

Read the following problems. We will not solve them here; our aim is to scrutinize them. See if you can detect a potential source for errors in each problem.

> A container in the shape of a right circular cylinder with radius 3 meters and height 11 meters is being filled with water at a constant rate of 40 cubic feet per minute. At what rate is the water level in the cylinder rising when the water level is 6 meters high?

A careful look will show that the units are different—the dimensions of the cylinder are given in meters, and the fill rate is given using cubic feet. One of these units must be converted before you begin working on the problem. Also, you may have noticed that the water level rises "at a constant rate," so the height is unnecessary to solve the problem.

> Nick has 100 feet of fence. He wants to use it to enclose a garden. What is the area of the largest garden he can enclose with 100 feet of fence?

Students often make the assumption that the garden is rectangular. In fact, a circular garden with a circumference of 100 feet will contain more area than any rectangular garden with a perimeter of 100 feet.

Just as you shouldn't make assumptions that are unwarranted, also be careful not to make generalizations too quickly. Examine the following problem:

> Compute the following: $(20 + 25)^2$, $(30 + 25)^2$, $(40 + 25)^2$

The answers to the first two expressions are 2025 and 3025, respectively. Study these two examples. Can you formulate a conjecture? Did you notice that the answer is composed of the four digits in the parentheses, in the same order? If you use this conjecture to compute $(40 + 25)^2$, you will make an error. The answer is not 4025 but rather 4225, and the answer does not follow the pattern. You shouldn't base a conjecture on too small a pool of trials. Don't be quick to generalize—scrutinize!

> Two competing pizza stores want to attract customers. One offers a free 10-inch-diameter pie with each regular pizza. The other has increased its pizza's diameter by 4 inches and not raised the price. Which store offers the better deal?

Careful scrutiny reveals that there is insufficient information to solve this problem. What is the original diameter of each store's regular pizza? What is the price per square inch of each deal, and how can it be computed without any prices being given? Sometimes advertisements are deliberately evasive so that consumers will have more difficulty making comparisons. The problem as posed has no solution.

> Two trains are riding on parallel tracks. One left Baltimore for New York City at 11 A.M., and the other left New York for Baltimore at 12 P.M. New York and Baltimore are 210 miles apart. The first train traveled at 60 mph, and the second traveled at 65 mph. At the time when the two conductors meet each other, which conductor is closer to New York?

It is astonishing to watch excellent math students work on this problem, applying all sorts of algebraic formulas, diagrams, variable representations, and so on. If you read the problem carefully and scrutinize it, you might notice that *when the conductors meet each other they are the same distance from New York!*

When scrutinizing, it is important not to detrimentally narrow the scope of your thoughts. Be open-minded about possible solutions. If you make an assumption, don't make it iron-clad. Possibly you are thinking along a track that will hinder your quest for the solution. The following problem illustrates this situation:

Find the rule that explains the following sequence: 8, 5, 4, 9, 1, 7, 6, 3, 2, 0

The solution will be given in this paragraph; if you want to pursue the solution on your own, stop reading now. Invariably, you will start by applying arithmetic rules to try to "coax" an algorithm that these numbers follow. Possibly there is one, although this author has never seen it. Open your mind. Think liberally. Don't think arithmetic—think spelling. The above sequence is the single-digit whole numbers arranged in alphabetical order according to their spelling! Our point is to warn you against considering only one road as you search for a solution. Expand your sights. You might need geometry to solve a probability problem, or calculus to solve an algebra problem. Always consider a broad range of options.

Although you might consider some of the examples here to be "trick" questions, they illustrate the need to scrutinize problems. Working on a question that you inferred rather than on the problem that was intended is not prudent. Solving problems can be hard enough without adding such "unforced errors." As your mathematical sophistication grows and you begin to tackle higher-level problems, careless reading can cause you to waste valuable time and effort. It is important to assess each problem by reading it carefully, critically, and fully.

Problem-Solving Strategies

We will discuss several problem-solving strategies in this section, but this overview could not possibly provide all the experience you need in problem solving. You'll need to draw on other experiences in order to become an accomplished problem solver. As mentioned earlier, you should read some of the recommended works on problem solving. Keep in mind that familiarity with problem-solving strategies will help you handle the new, nonroutine claims and conjectures you meet in your research. Included in the discussion of the strategies are some sample problems. If you would like to work through the problems before reading the solutions, stop reading after each problem is posed.

Draw a Diagram

Most students are in the habit of drawing diagrams in geometry problems, and rightfully so. A picture truly is worth a thousand words, and a diagram can help you discover a solution because the visual images enhance the verbal explanations. Diagrams can help solve nongeometrical problems as well. Let's see how drawing diagrams can help us solve the following problems.

> A worm is at the bottom of a 12-foot-deep well. Each day it crawls up 2 feet, but during the night it slips back 1 foot. In how many days will the worm get out of the well?

Solution: It seems that the worm gains 1 foot per day. This reasoning accounts for the popular but incorrect answer of twelve days. Let's draw a diagram and follow the worm's progress day by day. We will label 1-foot increments on the side of the well and draw on the diagram to follow the worm's progress.

As you can see, on the eleventh day, the worm is at the top and does not slip back! The answer is therefore eleven days, not twelve.

> There are six officers in a school's student government organization. A committee made up of two of them is to be randomly selected to represent the student government organization at local school board meetings. How many different committees of two are possible?

Solution: We can draw a circle and place the six people, represented by six letters of the alphabet, around the circle. Next we draw a line segment from each person to each of the other students. Each line segment represents one committee. Now we simply count the number of line segments to find the answer.

The answer is fifteen different committees of two. The different committees can be listed by reading each line segment in the picture.

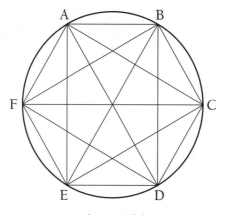

Make a Table

A table is a systematic way to list the different possible scenarios of a problem. Examination of a table can help you find the solution to a problem. Let's look at a problem posed previously in this chapter. Keep in mind that you might want to draw a diagram *and* use a table to find the solution, or use just one of the strategies.

> Eve is planning a rectangular flower garden that will contain a 36-square-foot flower bed. She wants to enclose the garden with a short decorative fence that is sold in 1-foot lengths that snap together. Each 1-foot length costs $3.99, plus 8% sales tax. Find the dimensions of the garden that requires the least amount of fence.

Solution: As mentioned previously, there is unnecessary information that can be eliminated from this problem. The task is to find the length and width that requires the least amount of fencing, not to find the cost of the fencing. (Eliminating unnecessary information is a problem-solving strategy in itself that is discussed later in this chapter.)

Let's set up a table of possible lengths and widths and the resulting perimeters. Remember that the length and width selections are not arbitrary; they must give an area of 36 square feet and must be in 1-foot increments.

A square measuring 6 feet on each side gives the minimum amount of fencing, 24 feet. The table organized the information and made it easier to analyze.

Length	Width	Perimeter
1	36	74
2	18	40
3	12	30
4	9	26
6	6	24

Use a Matrix

Closely linked to the "make a table" strategy is the "use a matrix" strategy. A rectangular array with rows and columns is called a **matrix**. Like diagrams and tables, matrices can help clarify information. The entries in a matrix can be symbols, numbers, or letters, depending on the particular problem. Let's see how a matrix can help us solve the following problem.

Janet, Rich, and Linda are students who have different hobbies and different pets. Janet's hobby is painting, and she does not have a bird. The student whose hobby is gardening does not have a dog. Linda's hobby is not gardening. The person whose hobby is skiing has a cat. Can you determine the hobby and pet for each person?

Solution: Scrutinizing this problem should tell you something. This story is too convoluted to try to decipher without a table. Underline the hobbies and pets—they will help us fill out the table. We will use initials to represent hobbies, people, and pets, the items to be compared. The strategy is to set up a matrix. There are different matrices we could use. Let's start with a simple matrix for matching pets and people:

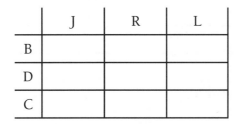

Some clues match hobbies with pets or hobbies with people, so we'll add two more matrices for matching hobbies:

	J	R	L	G	P	S
B						
D						
C						
G						
P						
S						

Number the sentences in the problem from 1 to 5. We will fill in an X or a ✔ in each box for no and yes, respectively. Copy a matrix like the one above on scrap paper and fill it in as we explain the solution.

Sentence 2 tells us that under Janet's name we should place an X next to bird (B) and a ✔ next to painting (P). Since Janet is the painter, Rich and Linda are not, so X's can be placed next to painting for Rich and Linda. Also, Janet is not the gardener or the skier. Sentence 3 tells us to put an X on the dog row under gardening. Sentence 4 tells us to place an X next to gardening under Linda. Now we see that Rich must garden and Linda must ski, so we place ✔'s in those boxes. Sentence 5 tells us to enter a ✔ in the cat row under skiing. As a result, the painter and the gardener receive X's in the cat row. Therefore, the gardener (Rich) must have a bird (✔). Finally, the painter (Janet) has a dog.

	J	R	L	G	P	S
B	X			✔		
D				X	✔	
C				X		✔
G	X	✔	X			
P	✔	X	X			
S	X		✔			

At this point, the rest of the chart could be filled in by correlating information from one part of the chart to another. But enough information is here already to allow us to state a solution: Janet is a painter and has a dog, Rich is a gardener and has a bird, and Linda is a skier and has a cat. How difficult would this problem have been without the matrix?

Solve a Simpler, Related Problem

Certain problems may overwhelm you due to large numbers and/or a large number of conditions. Solving a simpler, related problem affords you the chance to reason out a strategy that might extend to the original problem. There are several different ways to create a simpler, related problem. You might use a number that is easier to manipulate in place of a given number in the problem. You might replace a variable with a number, work out the problem with that number, and then apply the process you used to the variable in the problem. You can also eliminate unnecessary information and/or change some of the conditions of the problem. Try using the "solve a simpler, related problem" strategy to solve the following problem.

Find the difference between the circumference of the large circle below and the sum of the circumferences of the smaller circles. The diameters of the smaller circles are given in terms of x.

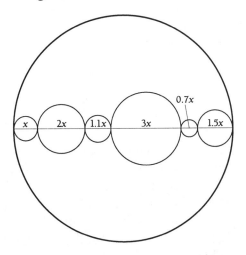

Solution: To analyze this problem, let's begin by simplifying it. Suppose we change the number of small circles and assign numbers to the variable diameters, relaxing the ratio conditions placed on the diameters of the small circles. These three changes create a simpler, related problem. We now solve the related problem.

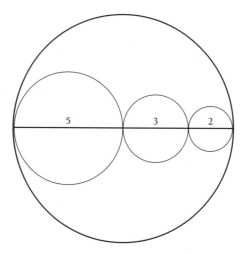

The diameter of the large circle is 10, so its circumference is 10π. The circumferences of the three smaller circles are 3π, 5π, and 2π. The sum of these circumferences is 10π, which equals the circumference of the larger circle. The difference is thus zero. Is this result a coincidence based on the numbers we picked? This is always a concern when you change a problem; however, in this case, you should pick other numbers and verify that the difference is still zero. This property can be shown algebraically. Often the answer to a numerical, related problem will get you interested in *attempting* an algebraic solution. You'll find this happening in the course of your math research.

Let's look at a more difficult example and its solution using the strategy of solving a simpler, related problem.

> How many consecutive zeros are at the right end of the number represented by 33!?

Solution: The number 33! is too large for your calculator. (What factorial *can* your calculator express without using scientific notation?) As you scrutinize the problem, you will realize that to try to find the product by repeated multiplication would be cumbersome and prone to error based on the sheer volume of computations. Underline the words *consecutive* and *right*. The plan is to work on solving the same problem with a smaller number, say, 12!, and see if we can extend the solution to 33!. Write out 12! as a product of consecutive numbers:

$$12\cdot11\cdot10\cdot9\cdot8\cdot7\cdot6\cdot5\cdot4\cdot3\cdot2\cdot1$$

The 10 should speak to you. The zeros we are looking for are formed by factors of 10. Does this mean that 12! has only one zero at the right? Think. Scrutinize. Think about the prime factors of 10, that is, 2 and 5. If each factor of 12! were factored into its prime factorization, we could look for sets of 2 and 5, knowing that each set will represent one zero at the end of 12!. Think more critically. There are many even numbers in the prime factorization, so there should be many factors of 2. A factor of 2 can create a zero only if it can be paired with the factor 5. The factors of 5 seem much more scarce. If we count the number of times 5 appears in the prime factorization and there are enough factors of 2 to pair them up, we can figure out the answer to the problem for 12!. The factors 10 and 5 each have one 5 in their prime factorization, so 12! has two 5s in its prime factorization. When these two 5s are paired with any of the even factors, two consecutive zeroes are created on the right. The number 12! thus has two consecutive zeros on its right end.

This strategy extends without much modification to 33!. Could you solve the problem with 33! without writing out the factors? Try it. See if you can verify that the solution is seven consecutive zeros.

Use Trial and Error (Guess and Check)

If there is a relatively small pool of possible answers to a problem, the trial and error method can be very effective. It is sometimes more prudent to try the few possible answers than to construct a lengthy theoretical argument to find the solution. Contrast the trial and error strategy with solving a simpler problem. We solve simpler, related problems when a complexity in the problem may hinder our ability to find a suitable strategy. Using trial and error is more suitable when a problem lacks complexity and has a small number of possible solutions we can try. The trial and error method often involves eliminating possibilities. If you can rule out some of the trials, you should. For this reason, we suggest that there is a difference between "guess and check" and "trial and error," even though the two expressions are often used synonymously. Hopefully, there will be some reasoning behind your trials; your trials shouldn't all be mere guesses, if possible. Try to solve the following problem using trial and error.

> In ten years, Nancy will be four times as old as Bernadette is now. Bernadette's age is now two years less than half of Nancy's age. How old will Bernadette be in twenty years?

Solution: Scrutinize—remember that ages are positive integers. Also, note that if in ten years Nancy will be four times as old as Bernadette is now, then guesses for Bernadette's age should be less than twenty-five for starters. We will try different ages, check each trial, and keep track of our trials in a table. We will use ages for "Bernadette now" as trials and enter them in column 1. Then we can fill in columns 2, 3, and 4 and check to see if the conditions of the problem are satisfied. If they are, columns 1 and 4 should match. We can rate our four guesses in fifth column.

Bernadette Now	Nancy in 10 Years	Nancy Now	2 Less Than Nancy's Age	Rating
10	40	30	13	Close
12	48	38	17	Colder
8	32	22	9	Very Close
7	28	18	7	Right on

Note that our second guess was farther off than the first, prompting a third guess in the other direction. Verify that the last line of the table satisfies the conditions of the problem. Bernadette will be twenty-seven years old in twenty years. Keep in mind that making a systematic list is often part of the trial and error strategy, because you need to keep track of the trials you have completed. You don't want to repeat any trials you have already tried, and you also want to allow yourself to see any possible pattern in the results.

Look for a Pattern

When an answer to a problem cannot be found directly, it is sometimes helpful to examine a sequence of solutions to simpler problems that lead up to the solution to the original problem. This strategy uses, in part, both the "solve a simpler problem" strategy and the "make a table" strategy because it is usually helpful to put the results in a table. Let's look for a pattern to solve the following problem.

Find the sum of the first hundred odd whole numbers.

Solution: As you read the problem, you will realize that you could write out the first hundred odd numbers and find the sum; however, there is a chance of error when you are adding so many numbers, even with a calculator. Let's look at the sum of a smaller number of odd whole numbers and see if we can gain any insight into the problem.

Look at the table on the next page. Do you notice a pattern? If n represents the number of odd whole numbers to be added, the sum is n^2. Try a few other examples to verify this conjecture. The table does not constitute an algebraic proof, but it gives you a reason to try to write a proof. (We discuss proofs in greater detail in Chapter 5.) The first hundred odd whole numbers have a sum of 100^2, which equals 10,000.

Number of Addends (n)	Expression for Sum	Sum
1	1	1
2	1 + 3	4
3	1 + 3 + 5	9
4	1 + 3 + 5 + 7	16
5	1 + 3 + 5 + 7 + 9	25

Use Algebra

The algebraic skills you have learned are very useful problem-solving tools. Let's use algebra to solve the age problem first posed in the trial and error section above. (We've already shown a solution using trial and error that included making a table.)

In ten years, Nancy will be four times as old as Bernadette is now. Bernadette's age is now two years less than half of Nancy's age. How old will Bernadette be in twenty years?

Solution: The reading step is crucial here, because you must first translate the words into algebraic expressions that accurately reflect the conditions of the problem.

Let b = Bernadette's present age.

Let n = Nancy's present age.

Underline the key phrases that you will use to create the equations. *Bernadette's age is now two years less than half Nancy's.*

$$b = (n/2) - 2$$

In ten years, Nancy will be four times as old as Bernadette is now.

$$n + 10 = 4b$$

If we substitute for b in the second equation, we have a linear equation in one variable:

$$n + 10 = 4[(n/2) - 2]$$
$$n + 10 = 2n - 8$$
$$n = 18$$

By substitution, $b = 7$. In twenty years, Bernadette will be twenty-seven years old.

Notice that you could have used the table from the trial and error section and filled it in with variables. Very often, the numerical manipulations required to fill in a table can help you determine variable representations for the problem. Let's add the variable representations to the previous table.

From the first two columns, we know that $4b = n + 10$. The first and last columns yield the equation $b = (n/2) - 2$. Notice that these are equivalent to the equations formed above and can be solved similarly.

Bernadette Now	Nancy in 10 Years	Nancy Now	2 Less Than Half Nancy's Age
10	40	30	13
12	48	38	17
8	32	22	9
7	28	18	7
b	$n + 10$	n	$(n/2) - 2$

When you complete a problem that you solved algebraically, verify that the answers fit the conditions of the problem. Why can't you simply substitute the answers into the equations to check the solution? Keep in mind that the most difficult part of using algebra is setting up the correct expressions and equations. Read and scrutinize carefully.

Work Backward

Did you ever know the answer to a problem and use it to figure out how you could do the problem? If so, then you have already worked backward to find a solution. Working backward is helpful when an end result is known and when you know the steps for getting from the unknown to the end result. Let's look at an example.

A furniture clearance center adjusts its prices weekly according to the following markdown schedule:

week 1 price (W_1)—price is as marked
week 2 price (W_2)—10% off week 1 price
week 3 price (W_3)—20% off week 2 price
final price (F)—25% off week 3 price
Beth bought a couch that had been in the store for five weeks for $324.
What was the original marked price of this couch during week 1?

Solution: Since we know the end result (the price Beth paid) and we know the rules used to arrive at it (the markdown schedule above), working backward to find the original price seems like the appropriate strategy. Since the couch was at the store for more than four weeks, let's work backward from the final price of $324.

The final price of $324 is 25% off W_3. Therefore, it is 75% of W_3.
   $0.75W_3 = 324$, so $W_3 = \$432$
The week 3 price of $432 is 20% off W_2. Therefore, it is 80% of W_2.
   $0.80W_2 = 432$, so $W_2 = \$540$
The week 2 price of $540 is 10% off W_1. Therefore, it is 90% of W_1.
   $0.90W_1 = 540$, so $W_1 = 600$
The original marked price of the couch at the clearance center was $600.

You probably have noticed that some of the solutions above used more than one strategy. Many solutions will be comprised of combinations of the strategies discussed. How will you know which strategies to use? Reread Polya's statement that opened this chapter. It is a SUPERB commentary on attaining problem-solving proficiency.

A Look at Looking Back

The last part of solving any problem is to read the problem one more time to make sure you answered the question. This is part of reviewing—looking back and checking your work. Regardless of what problem-solving technique you employ, it is always a good idea to see if the answer you find makes sense. Let's review the age problem presented earlier in this chapter.

> In ten years, Nancy will be four times as old as Bernadette is now. Bernadette's age is now two years less than half of Nancy's age. How old will Bernadette be in twenty years?

The answer is that Bernadette will be twenty-seven. The specific method of solution is not our focus here. Before checking the answer against the conditions of the problem, we want to see if the answer makes sense. First, the age of twenty-seven for a human being makes sense. An answer of less than twenty years would not make sense. (If Bernadette is alive now, in twenty years she must be at least twenty.) We would also probably be uncomfortable with an answer of 347 years old, which could be arrived at as a result of an error. After examining this level of "Does the answer make sense?" we can check the answer against the conditions of the problem to see if it satisfies them.

Sometimes it is more difficult to determine whether an answer makes sense. You may not have a frame of reference, as you did with the age problem. For example, suppose you did a report on the Empire State Building, a 102-story skyscraper in New York City. How many miles of electrical wiring are in the Empire State Building? Three? Sixty? Four hundred? Two thousand? It's difficult to have a handle on whether a particular answer makes sense in this case. With this in mind, read through the famous "Birthday Problem."

> How many randomly selected people would you have to assemble to make the probability of two or more people having the same birthday (month and day) greater than 50%?

There are 366 possible birthdays, including February 29. Scrutinize. The answer can't be greater than 367 people, because we are guaranteed a match with 367 people. How many people do you think it would take to get the probability over 50%, where there is a better chance of a match than no match? More than 180? Fewer? The answer is 23. The probability of at least two people having the same birthday is greater than 50% if there are just twenty-three random people polled. Try this in school. Check your classes that have more than twenty-three students. You'll be surprised at how few people it takes to create this matching birthday situation! The mathematical solution requires a background in counting principles of probability. If you have learned about permutations, combinations, and the probability of a complement, you can attempt the solution.

We discuss this here because the correct solution seems unreasonable. Most people think 23 is too small a number, that it's incorrect. The theoretical mathematics make the solution clear, and if you experiment throughout your classes (many classes), you will believe it is correct. This is an example of a situation when "looking back" requires you to examine your strategies and calculations for correctness, because the answer may not seem to make sense.

Try this popular circumference problem.

> A steel band is placed around the earth, snugly fit at the equator. (The equator is approximately 25,000 miles in circumference.) The band is cut, and a 36-inch piece of string is spliced into the steel band. This new circular band is placed around the earth, centered off the earth's surface, so its center coincides with the center of the earth. A gap is created between the equator and this circular band. What could you fit in this gap? A hair? An index card? A computer

diskette? How wide is this gap?

The gap is approximately 6 inches wide! In fact, the gap is the same width as the gap created if 36 inches of string are added to a band placed around a *nickel!* How could 36 inches have such a profound effect over 25,000 *miles?* How could they have the same effect on the relatively tiny nickel's circumference? You can take some string and do the experiment around a trash-can cover and a nickel to verify the result. Mathematically, the solution is not complicated.

$$C = 2\pi r$$

If the circumference increases 36 inches, the radius must increase. Let's call the increase in the radius g, for gap. The new, larger radius is therefore $r + g$.

$$C + 36 = 2\pi(r + g)$$
$$C + 36 = 2\pi r + 2\pi g$$

Since $C = 2\pi r$, we can subtract these equal quantities from both sides of the equation.

$$C + 36 - C = 2\pi r + 2\pi g - 2\pi r$$
$$36 = 2\pi g$$

Use an approximation for π: $\pi = 3$.

$$36 = 6g$$
$$6 = g$$

The gap is approximately 6 inches.

The formula shows why the original circumference does not affect the answer; hence, the nickel and the earth yield the same gaps. Drawing a graph also shows why this answer makes sense. Since $C = 2\pi r$, we can write the following:

$$r = (1/2\pi)C$$

The graph of this equation on the r, C axes is a straight line with slope $1/2\pi$ and y-intercept zero. The graph shows that any 36-inch increase in circumference causes the same (approximately 6-inch) increase in the radius, regardless of the original circumference.

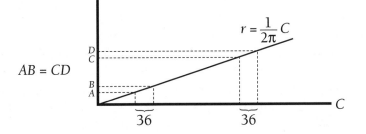

Try this problem on your friends and family. The answer is so amazing, don't be surprised if people tell you your solution is wrong! Again we see that correct answers sometimes aren't what we expected, and we may think they don't make sense. In such cases, we must follow Polya's steps and check all the strategies and calculations. The following problem is another classic with an unexpected result.

> Three people pay $10 each (total $30) to check into a hotel. The manager realizes that they were overcharged and gives $5 to the bellboy to return to the three people. The bellboy decides to return $1 to each of the three people and keeps $2 for himself. The three people paid $9 each, for a total of $27. The bellboy received $2. What happened to the other dollar?

We leave this problem for you to scrutinize and solve.

Problems and Questions

Did you ever wonder why we *park* on a *drive*way and *drive* on a *park*way? Why is it that when you send something by *car* it's a *ship*ment and when you send something by *ship* it's *cargo*?

Think back about what you've read in this book. Chapter 1 focused on questions, and Chapter 2 focused on strategies to find answers. Notice how many questions were asked and how many were inferred. Problems inspire questions and these questions can become new problems. In carrying out thorough mathematics research, you will face new questions that arise from your work. As mentioned in the build step, you should write down these questions. You may decide to answer them as part of your research. Practice writing questions. See how many questions you can create based on any of the problems in Chapters 1 and 2. Asking questions is a sign that you are thinking open-mindedly—you are entertaining other scenarios. Good, pertinent questions are an indicator that you understand the problem. There is an art to formulating a clear question that has well-defined conditions. Practice this as you problem-solve and as you do your research. For further insight into asking questions, read *The Art of Problem Posing*, by Stephen Brown and Marion Walter.

Try writing your own problems. You can use an original extension of a given problem to create a new problem or think of some problems either on your own or cooperatively with your classmates. Ask your math teacher, department chairperson, or math team coach to act as an editor and reviewer for your problems. You might want to have your math class start a problem-solving column in your school's newspaper, using your original problems. Maybe you can solicit a local ice cream or pizza parlor to sponsor a monthly prize for the winner of a newspaper problem-solving contest. Your class can publish its own problem-solving booklet, using problems created by the students, to be distributed to students, teachers, and administrators.

The National Council of Teachers of Mathematics (NCTM) publishes a journal called *Mathematics Teacher*. Each month, a calendar features a different original problem for each day of that month. Try sending some of your class's original problems, with solutions, to the NCTM for possible publication in the calendar:

National Council of Teachers of Mathematics
ATT: Calendar Problems
1906 Association Drive
Reston, VA 22091-1593

It is truly exciting to see your name and your problem in print—a problem that will be read by people all over the United States and Canada!

Be Determined

Musicians practice. Athletes practice. Mathematicians practice. Sometimes, a difficult challenge leads to frustration. Perhaps you found some of the problems in this chapter frustrating. Good researchers and problem solvers build confidence and perseverance based on the fact that their previous problem-solving persistence led them to correct solutions. You'll need much more problem-solving practice than this chapter can provide. You should seek out books whose primary focus is problem solving. Problem-solving and brainteaser books are readily available in bookstores. You may want to learn how to play mathematical games to sharpen your insight and build your concentration and persistence levels. Some excellent games are available at bookstores and hobby stores. React constructively to the challenges you face; the teasing and frustration of a seemingly unsolvable problem should trigger determination. Ask questions. Form groups of students that can discuss the problem and exchange or pool ideas.

Don't set a time limit on finding a solution—it's not realistic. Sometimes you will have to leave a problem and come back to it later; therein lies the reason for keeping legible notes. Frequently, when you leave a problem, an idea for a solution will come to you at a time when you are engaged in some other activity, because all the while the problem was on your mind. You may give up on a problem, *or think you have given up*, only to have a solution come to you when you are playing softball, listening to music, or walking to school. At that point, you've reached the epitome of the spirit of problem solving; you'll be able to apply your problem-solving experience to the problems you encounter in the course of doing mathematics research.

Chapter Three
Writing Mathematics

When you complete your mathematics research, you will write a formal research paper in which you communicate your work to others. In preparation for this, we begin discussing writing and communicating in mathematics. By the time you begin your research paper, you should understand what it takes to produce high-quality written mathematics. Most of your research will be new mathematics for you. You will need to do an effective job of communicating it to your readers. How well can you communicate the mathematics you already know? Let's try an experiment adapted from Richard Sgroi's article in the February 1990 issue of *Arithmetic Teacher*.

You'll need a partner for this experiment. You are going to create a geometrical sketch and try to have your partner draw it without seeing it. A sample is shown below.

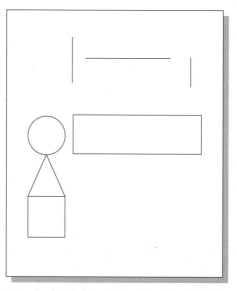

Sample sketch for communication activity.

Your partner should have the same size paper as you use and a ruler. Sit back-to-back. Use your mathematics vocabulary to describe the sketch you have created to your partner, one step at a time, with pinpoint accuracy. Your partner cannot ask any questions. When your partner's sketch is completed, superimpose one paper on the other in front of a light source and see how well they match. Discuss with your partner places where miscommunication caused the drawings to differ. Were the errors in the description or in the interpretation? Switch roles and try the experiment again. Make a list of mathematics vocabulary that can be used to describe geometrical sketches. You should be able to think of dozens of words right off the top of your head.

As you can see, a comprehensive explanation using precise language is the cornerstone of this activity and mathematics communication in general. It is also the cornerstone of your writing. Your research paper will require you to communicate all your research findings in writing. Writing is not new to you. You have been writing since elementary school. From one-sentence answers to questions, to full-paragraph explanations, to English compositions and history papers, you have done lots of writing. Is writing mathematics different from your other writing experiences?

To see just one way in which it differs, consider this question:

Do you think you could distinguish between a math book and a novel, each opened to a random page, from 30 feet away without being able to actually read the print?

Look at the page layouts below. Which page looks as if it is describing mathematics?

Why does the mathematics page (on the right) look so different? Is there a purpose to the broken-up page layout? Mathematics writing has some idiosyncrasies that you may not have employed in your previous written work. This chapter will help you incorporate the writing skills you already possess with some suggestions that will help make your mathematical explanations as effective as possible. Acquisition of these skills will make you a more versatile writer. As with all writing, you can master mathematics writing skills by practicing. Chapter 4 offers you an activity, called the Math Annotation Project, that will help you practice. By the time you start writing your research paper, you'll already have valuable experience writing mathematics. This will make the research-paper writing process smoother and the writing itself more polished.

How to Write Mathematics

Perhaps you have done some informal mathematics writing in your math classes. You may have explained solutions or problem-solving plans or written reflectively. Reflective pieces allow you to examine the work you've done and discuss how you did it and how it could be improved. Introspection is a valuable experience that can help you when you meet future mathematical challenges. If you are interested in informal mathematics writing, you should read *Writing to Learn Mathematics* by Joan Countryman. The Math Annotation Project and your math research paper, however, will not be informal, reflective pieces but attempts to technically describe mathematics in a clear, comprehensive manner.

All the skills you use to write essays will be incorporated into your mathematics writing. You'll need to tap what you've learned about sentence and paragraph construction, punctuation, and grammar. You can use a writing handbook, available from your library or English department, as a reference. In this chapter, we will discuss writing tips, formatting tips, and finally proofreading and editing tips. Be sure to read this chapter in its entirety before you begin the writing activity in Chapter 4. As you read, imagine how you could incorporate the tips if you decided to rewrite the notes you took in a recent math class. You could actually rewrite a portion of some homework or classwork using the tips. If you do, analyze the finished product and note how using the suggestions in this chapter improved it. Incorporate some of the tips as you take notes during future math classes.

Writing Tips

❧ Before you write, you need to create an outline. In writing your research paper, you will have to make many decisions on how to organize it. Chapters 4 and 8 discuss outlines. Outlines can be written by hand and can be revised as necessary.

❧ The writing of the formal paper should be done on a word processor. (Journal notes are usually done by hand.) Writing by hand or typing is terribly inefficient when the work will be undergoing an extensive, repetitive editing process. Good, readily accessible word-processing programs specifically designed for writing mathematics are available. Talk to your school's computer teachers about what software you should use. Once you start writing, make sure you have two different disk copies and hard-copy (paper-version) backups of your drafts.

❧ From the very first draft you create, number the pages. You might want to include a header or footer that places your name at the top or bottom of every page, in case pages are ever lost.

❧ Double-space all drafts and the final paper. Double-spacing leaves room for you, the writer, and any student editor or teacher editor to insert comments. See the edited sample Math Annotation Project in Chapter 4 and the student samples in Appendix B for examples.

❧ Don't start a sentence with a symbol.
Incorrect: "c is the length of the triangle's hypotenuse."
Correct: "The length of the triangle's hypotenuse is c."

❧ You may notice in your math textbooks that variables are written in italics. If you have the time, you may decide to add a professional touch by italicizing your variables. This can be done when the written work is completed.

❧ When you discuss the data in a table, you might be inclined to start some sentences with a numeral. Do not begin a sentence with a numeral. Either word the sentence or spell out the number.
Incorrect: "112 is the eighth entry in Table 3, column 2."
Incorrect: "There are infinitely many prime numbers. 2 is the only even prime."
Correct: "The eighth entry in Table 3, column 2, is 112."
Correct: "There are infinitely many prime numbers. Two is the only even prime."

❧ Use correct, sophisticated terminology.
Incorrect: "Next, plug in 5 for x."
Correct: "Next, substitute 5 for x."
  Suppose you need to discuss numbers in the equation $4x^2 + 4y^2 + 36x + 24y - 100 = 0$.
Incorrect: "The number next to the x is 36."
Correct: "The coefficient of x is 36."
  Suppose that in another step you decide to write the above equation as
$4x^2 + 36x + 4y^2 + 24y - 100 = 0$.

Incorrect: "Things in the equation were moved around."
Correct: "The terms $36x$ and $4y^2$ were commuted."

❧ Don't cram excessive amounts of material into a single diagram. Math textbooks and journal articles often do this because of space limitations. Often a single diagram is used to illustrate several major points that could be better explained in several less complex diagrams. Even if a complex diagram is necessary to make a point, the diagram can be built in several stages, with each stage gradually more complex than the one before it. An example of a graduated series of diagrams can be seen in Jeff's work in Appendix B_4. When Jeff needed to show how any rational multiple of a given line segment could be constructed using compass and straightedge, he set up a series of diagrams that made the steps clear for the reader. Jeff describes the construction of two-thirds of a given line segment with length a.

❧ If you are in a class or school in which several students are doing annotation projects on the school's computers, your teacher may decide to organize a central file composed of all the diagrams the students create on the computer. You can save your diagrams in this diagram bank (as well as in your own file) and make your diagrams available to others. You may benefit by "withdrawing" a diagram you need from the file. Over the years, the file will grow, and you can call up circles, trapezoids, right triangles, and so on with ease. Be sure to give credit to the source of any diagram.

❧ Watch the use of pronouns. It is better to repeat a term often than to replace the term with *it* or *they* if there is a chance the reader will get confused. In the following passage, the word the pronoun *it* refers to is clear.
   "The cube root of two cannot be expressed as the quotient of two integers. Therefore, it is not a rational number."
   This word the pronoun refers to is called the **antecedent** of the pronoun. In this case, the antecedent of *it* is "cube root of 2." You may think you are being too repetitive when you use a word over and over again. Remember that clarity is your goal. The material you are presenting is challenging enough for the readers; don't confuse them unnecessarily. Read the following potentially misleading uses of pronouns.
   "The students loved the books, but they were boring." Were the books boring, or were the students boring? This question has tremendous impact on any discussion that follows. You don't want readers to make incorrect inferences about the rigorous mathematics in your paper.
   "Rational numbers can be expressed as the quotient of two integers. Every one of them can be expressed as a terminating or repeating decimal." Does *them* refer to *rational numbers* or to *integers*?

Sometimes the quirks of grammatical usage can be confusing, even if they don't involve pronouns, antecedents, or wrong words. Read the following passages for some examples.

A corporate slogan: "Acme—Building Machinery Since 1933." Has Acme been building *machinery* for varied uses since 1933, or has Acme been producing *building machinery*, which specifically says that they make machines used for building? The exact meaning isn't clear.

A radio disc jockey's comment: "Probably the best song ever by the Beatles." Was the song the disc jockey was referring to the best *Beatle*-song ever, or the best *song* ever, which happened to be by the Beatles? Do you think a comma changes the meaning, that is, "Probably the best song ever, by the Beatles"?

A newspaper article's title: "Endangered Wildlife Cards." Is there going to be a series of cards strictly about *endangered* species, or is a current series of cards about *all types* of wildlife going to be discontinued due to limited sales (that is, the series is endangered)?

A clown rental agency: "You know our commitment to fun shows." Is *fun* a noun or an adjective? Is *shows* a noun or a verb? Is the agency saying its commitment to fun is easy to see, or that they have a commitment to shows that are fun?

Usage problems such as these are more apt to be spotted by a proofreader than by you, the author. You are very familiar with the work and, as a result, know the intention of potentially misleading phrases. Try to remain aware of such phrases as you write.

⋙ Avoid passive verb constructions and use active verbs instead.
Passive: "The last conjecture was made by Joshua."
Active: "Joshua made the last conjecture."

⋙ Don't be judgmental. Phrases like "It is easy to see that . . . ," "It is clear that . . . ," or "Simply . . ." should not be used. Something obvious to the author may not be obvious to the reader.

The writing tips given above apply to all types of writing, but you know from experience that mathematics writing has a distinctive "look." The following section will help you set up your mathematics writing in a professional format.

Formatting Tips

❧ Mathematical expressions that need to be "digested" by the reader before continuing should be centered on their own line. Examples can be found in any mathematics textbook. See Appendix B for additional examples. There is a constant need to draw attention to major steps and expressions central to the development of a topic. Do not bury these expressions in the midst of a paragraph. Bring them to center stage. Let's look at an example.

> To find the slope of the line $3x + 4y = 24$, you should change the equation into the slope-intercept form of the line, $y = mx + b$. Subtract $3x$ from both sides of the equation. The result is $4y = -3x + 24$. Next, divide both sides of the equation by 4. The result is $y = -3/4x + 6$. The coefficient of x, $-3/4$, is the slope of the line.

Compare the paragraph above to the following:

To find the slope of the line

$$3x + 4y = 24$$

change the equation into slope-intercept form:

$$y = mx + b.$$

Subtract $3x$ from both sides of the equation and then divide both sides by 4:

$$\frac{4y}{4} \begin{array}{c} 3x + 4y = 24 \\ \underline{-3x \qquad = -3x} \\ \end{array} \frac{-3x + 24}{4}$$

$$y = (-3/4)x + 6, \text{ so the slope is } -3/4.$$

Centering a mathematical expression and highlighting it (putting it on a line by itself) allow the reader to follow the development more easily. Keep this in mind as you write and edit. As you examine the examples of student work in Appendix B, note the centered, highlighted expressions.

❧ Use headings to divide your paper into different sections at logical junctures. Devise a hierarchy as to how you will indicate new sections and divisions of individual sections. For an example, let's look at a subdivision of a mathematical paper on the real numbers. Imagine that there are several pages of text in between each heading.

THE COMPLEX NUMBERS

THE REAL NUMBER SYSTEM

The Rational Numbers

The Set of Integers

The Set of Counting Numbers

The Irrational Numbers

PURE IMAGINARY NUMBERS

Note how the headings show that the material on integers is a subdivision of the material on rational numbers while the information on rationals and irrationals is mutually exclusive. The material on counting numbers is a subdivision of the section on integers. The material on pure imaginary numbers is not a subdivision of the real number section or any of its subsections. You will learn more about headings in Chapter 8.

❧ Number all tables and write a caption for each table. The heading goes on top of the table and is underlined. Once the table has been numbered, you can refer to it in your text. You may decide to single-space numerical entries in a table. Do not break a page in the middle of a table. If, as part of a revision, you insert a new table, some table numbers and references to table numbers will need to be changed. An example of a table is shown below. See Appendix B for more examples.

Table 3 Base 10 to Base (-3) Equivalencies

Positive Integers		Negative Integers	
Base 10	Base (-3)	Base 10	Base (-3)
1	1	-1	12
2	2	-2	11
3	120	-3	10
4	121	-4	22
5	122	-5	21
6	110	-6	20
7	111	-7	1202
8	112	-8	1201
9	100	-9	1200
10	101	-10	1212

❧ Number all figures (illustrations, graphs, diagrams and photographs) in a separate numbering system from the tables (Figure 1 and Table 1, for example, are independent of each other). Figures can be drawn by hand or on a computer. Write a caption for each figure. Center the caption underneath the figure. Once the figure has been numbered, you can refer to it in the text. If, as part of a revision, you insert a new figure, some figure numbers and references to figure numbers will need to be changed. Below is Figure 00 from Robin's paper on tangent circles in triangles. Robin numbered all of her figures zeros until her paper was finished. Then she inserted all of the correct numbers. See Appendix B for more examples of figures and captions.

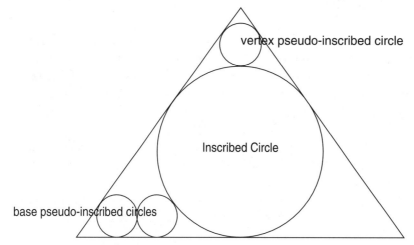

Figure 00: Illustration of Inscribed Circle and Pseudo-Inscribed Circles

Proofreading and Editing

Although other people may proofread and edit your work, you should proofread all drafts of your written work before you submit them.

Proofreading Tips

Proofreading involves reading your written work and looking for language-specific problems. You have probably been proofreading your school papers for years. When you proofread, you are looking for typographical errors, incorrect punctuation, spelling errors, inconsistency of symbols, incorrect numbering of figures and tables, and incorrect word usage. Your paper will be double-spaced, allowing you to make corrections in the spaces between the lines.

&✒ You can use symbols as notes to yourself when you proofread. Your teacher may use symbols on the drafts you submit for comments. You can also use symbols if another student asks you to proofread his or her paper. The following is a list of selected symbols commonly used by proofreaders. Acquaint yourself with these symbols.

Symbol	Translation	Example
⟃	move text	because (in this equation) there are no constants.
ℓ	delete	~~As~~ shown in Fig. 7
◡	delete space	The inter section of the sets . . .
⊙	insert period	is $x = 3$⊙There is one root.
⋏	insert comma	is $x = 3$⋏but this is not . . .
word ⋏	insert word	there are ⋏many prime numbers. *infinitely*
∩	transpose letters	Pythagorⱡⱥn Theorem
∪	transpose text	prove carefully the theorem
/¶	start new paragraph	is true./The next proof will . . .
≡	capitalize a letter	fermat's famous theorem
/	use a lower case letter	thinks Algebra is fun.
___ ⟨ital⟩	put in italics	did his own research ⟨ital⟩
∿ bf	put in bold print	is called a matrix. ⟨bf⟩
……⟨stet⟩	ignore the proofreading marks	is ~~quick and~~ simple proof. ⟨stet⟩
◯	don't abbreviate	likes Euler's (Thm) the best.
◯ sp	spelling error	the (Fibonaci) Sequence sp

These shorthand symbols take care of many language-specific corrections. You can create your own symbols to make notes to yourself as you proofread. Each time you read your work, check for language-specific errors by proofreading carefully.

≈ When papers are spell-checked by a computer, only misspelled words are found. There are two things to watch out for when you spell-check a mathematics paper. First, be aware that the computer may highlight algebraic expressions as misspelled words. Also, the computer will not be able to find wrong words. Look at the following examples: "The difference between six and five is on." The writer meant to say "one" instead of "on"; a computer spell-check would not find this error. "If the hypotenuse is even, hen the triple is not primitive." The writer meant to say "*then* the triple . . .," but the incorrect word, *hen,* is actually an unintended word that is spelled correctly. In such cases, the world's oldest spell-checker—the human brain—is necessary.

Editing and Revising Your Paper

Editing is subject-specific. When you read your paper, you will need to check the *presentation*—the style, flow, and organization of the paper. Are there redundancies? Should some paragraphs be moved? You will also need to check the *mathematics*. Make sure it is presented in a logical order with sufficient explanation, diagrams, tables, examples, proofs, counterexamples, and so on. You will need to check that the letters used to represent points, lines, and variables are consistent. Be sure you check even simple algebra and arithmetic.

As you complete pages, proofread and edit them yourself until each draft is as polished as you can make it. Then submit it to your teacher for comments and suggestions. When it is returned to you, make all the teacher's recommended corrections. Some could be as small as punctuation corrections, while others could be major revisions. You may not understand some of the comments, and you might have a reason for not wanting to make certain changes. Discuss these aspects of the editing with your teacher. Continue writing the next part of your paper. Proofread and edit, and submit another draft. You can include your revisions of previously edited versions with the subsequent drafts of material. Keep all your drafts after they are edited. Continue this process until the entire paper is complete.

The continuing process of writing drafts, proofreading and editing them, submitting them to your teacher, and revising them serves to improve the content and flow of your work. This cycle works effectively whether the writing is for newspapers, magazines, television, advertisements, instruction manuals, textbooks, song lyrics, poetry, annotation projects, or research papers. This is how writing gets fine-tuned. Papers written in haste do not exhibit the polish of a refined paper that has undergone extensive editing and proofreading. Proofread and edit as you convert your fragmented scratch-work (typical of mathematics) into complete sentences and paragraphs. Proofread your computer screen frequently during word-processing sessions, then proofread again when you print out your work. Everybody makes mistakes. Good proofreading is an indication of care, effort, pride, and quality.

Your teacher can help you with the editing process. You can have students who are members of the target audience proofread your work. They can serve as a good barometer of how well you are getting your point across. You may even want someone unfamiliar with your topic to proofread your work. If a reader outside of your target audience thinks the piece is clear, then your target audience certainly will.

Common Errors

Improving your writing requires practice and instruction, and all comments made are designed to help you produce a better written piece. All drafts will receive some constructive suggestions when read by your teacher. Try not to make errors that could have been avoided by using a writing handbook, the suggestions in this chapter, or the list below of teachers' comments that seem to surface frequently on student drafts.

ã "Number the pages": You need to number the pages so you and your editor can refer to them when making comments. Often students don't number their drafts, probably because they start submitting them when they are only a few pages long.

ã "New section heading needed here": Parts of papers need to be logically divided into separate sections, with appropriate headings.

ã "Center and highlight": Long, consecutive full paragraphs will be a rarity in a mathematics piece. Crucial math points are often centered and highlighted so they can be mentally digested by the reader before the reading continues.

ã "Segue needed here": As you divide your project into separate sections, include a logical connective that bridges the section that is ending to the next section. A transition question, statement, or paragraph that motivates the need for the next section makes the paper flow sensibly. These connections are called segues.

ã "Proofread": This comment is written when an editor finds mistakes that were avoidable, from typographical errors to word-processing mistakes. As text is moved around on a word processor, cut and pasted, copied, edited, and so on, proofreading becomes a more critical step in the revision process, not less, as some people think. Sometimes blocks of text appear in two places, or sentences are broken up because some editing took place. These must be caught by proofreading.

ã "Not a sentence": You may write phrases that are punctuated as sentences and are not grammatically complete. These should be noted and corrected before the draft is submitted.

❧ "I don't understand": The editor will read your drafts as someone who is unfamiliar with the content. Most often this is a charge for you to do a better job of explaining a particular point. Sometimes a person who is very familiar with something assumes the reader shares some of this knowledge. Consequently, the material is not explained in complete detail. (For example, did you ever give someone directions to your house and leave out a crucial part of the directions because the route is so familiar to you?) When you edit, try to empathize with your reader—read your work as someone who is not familiar with the material to determine whether the passage is clear and precise. Pretend your project is going to be read by a student who was absent and is relying on you for a comprehensive explanation.

❧ "Give an example": If a claim is made, often a mathematical example can complete the job of relaying a message that began with a verbal explanation. When algebraic statements are made, it often helps to give a numerical example. The following conjecture is stated in algebraic terms:

If $a \mid c$ and $b \mid c$, and a and b are different prime numbers, then $ab \mid c$.

To help your readers understand it, you could include a numerical example. Let $a = 3$ and $b = 2$. Make a list of numbers divisible by 3. Make another list of numbers divisible by 2. Circle the numbers that are divisible by both 2 and 3, and point out to the reader that they are all divisible by 6, which is the product of 2 and 3. Now, when you try to prove the conjecture, its meaning should be clearer to your readers.

❧ "Reword/awkward": Sometimes text that is technically correct can be shuffled and altered to read better. Avoid run-on sentences. Use a writing handbook as a reference throughout your math writing projects.

This chapter has given you suggestions designed to help you write effective mathematics. You will need to keep these points in mind as you write your research paper. You should also keep these points in mind when you make entries into your research journal, which will be discussed in Chapter 7. Chapter 4 describes an activity that will help you practice incorporating these ideas into your mathematics writing: the Math Annotation Project.

Chapter Four
The Math Annotation Project

Chapter 3 introduced some writing suggestions that you will use in writing your research paper. You should practice these skills, because they will help you understand and communicate all of your mathematics work. You can become proficient in mathematics writing by practicing and critiquing your work, and you should gain some writing experience before it is time to write your research paper. The Math Annotation Project's purpose is to improve writing and note taking, skills essential for good research. An annotation is an explanatory note designed to help the reader understand a passage. Students who intend to write research papers should complete this preparatory assignment before writing their research paper, but the Math Annotation Project is a solid learning experience for any student in any mathematics class.

You are about to become an author. This project will teach you how to rewrite the notes for a mathematics topic you learned in school. The topic could be as short as a one-day lesson or as long as a three-week unit. Your class notes form the skeleton of the project. Notes taken hastily during class discussions are usually incomplete fragments of information that can be greatly enhanced by annotation. As you examine your notebook, you might notice a lack of explanatory material. Would you be able to use your notes effectively to understand a concept months later? How easy would it be for absentee students to learn from the notes they copy from fellow students? Class discussions often don't involve complete-thought sentences. The annotation process requires you to review and rethink the topic. Each idea covered in your notes must be explained in full sentences. To be able to write such sentences, you must understand the material and learn how to explain it. Because you are writing about mathematics material you have already learned, you can concentrate on the writing and explaining and not on learning new mathematics. A sample annotation project is included in this chapter. Appendix B_{10} shows examples of pages from students' annotation projects. Take a look at them.

The tasks you must complete for your annotation project are given here in chronological order.

❏ Create an outline. The notes you took in class are the skeleton of your outline. In the annotation project, you use the exact notes taken in class, with the same examples, derivations, and so on. Since you are enhancing these notes, your outline will follow the presentation of the notes, with any extras you supply appropriately placed. If you are annotating only one day's class notes, your outline will not be long. You can revise your outline if, after working on the project, you decide to change the order of presentation of material or add certain features to your project. (Outlining a research paper from scratch, which requires many more decisions, is discussed in Chapter 8.)

❏ Staple the pages of each draft and the pages of the final project with one staple in each upper left-hand corner. Do not use plastic report covers; these make it harder for the reader to handle, especially if comments have to be made in the margins.

❏ Create graphics by hand or by computer.

❏ Add cartoon characters, symbols, computer programs, poems, songs, stickers, logos, mnemonics, and so on to help the reader understand the math concept being cov ered. It is imperative that no feature of your writing handicap the main goal of clarity.

❏ Add original examples, examples from a textbook (properly cited), original ques tions, and other related material.

❏ Add diagrams, captions, tables, examples, counterexamples, warnings, reminders, and so on, to aid your explanations.

❏ Offer problems both with and without solutions.

❏ Give in-depth solutions to the unit test that was given on your topic. An analysis of the test can include alternate solutions, time-saving strategies, and original ques tions modeled after the test questions.

❏ Create a cover page. The cover page is straightforward. It does not need borders, artwork, colors or pasteups. The cover page should contain essential information such as the title of your project, your name, grade, school, and date. Appendix B_1 shows a sample cover page for a research paper; the cover page for an annotation project is basically the same. The title is written in uppercase letters, in inverted pyramid form.

❏ Create a list of new terminology used in your work. The first page should be a list of key terms covered in the unit with the page number of their introduction. This list can be compiled when all of the written work is completed.

You will complete the annotation project with the help of a teacher who acts as your editor. You will submit drafts of the project each time you complete a few pages. Your teacher will correct these pages, and your next draft can include a revision of these pages, along with new material. Don't write an entire project without submitting drafts. You'll learn more about writing mathematics as you read and respond to the teacher comments. The final project will be of a high quality because you will have incorporated all of the comments. Periodically, you may want to do an annotation project on a single day's class notes to get more practice. This will improve both your understanding of the mathematics you are learn-ing in class and your writing. To give you a better idea of what an annotation project might look like, a sample excerpt of class notes and the accompanying annotation project follows.

A Sample Annotation Project

This sample annotation project represents only one part of a one-day lesson. We will examine the notes, an outline, a first draft with editing, and the final Math Annotation Project. A portion of class notes taken in a geometry class is shown below. Imagine the class discussion that took place as the lesson developed. Notice that much of this commentary was not written on the board and consequently did not end up in the notebook.

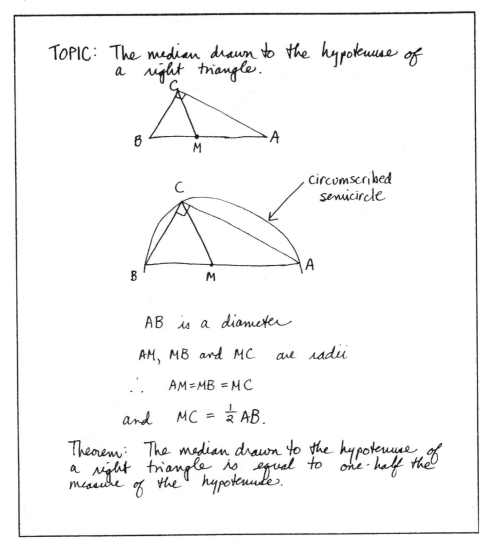

Sample of a student's notes.

The outline shown below covers the material in the notes and includes some enhancement. The outline features constructions and a discovery activity, as well as a discussion of the classwork.

ANNOTATION PROJECT OUTLINE

I. State topic

II. Three constructions
  A. Right triangle ABC
  B. Perpendicular bisector of hypotenuse to get midpoint of hypotenuse
  C. Median to hypotenuse

III. Discovery activity
  A. Give three more right triangles with median drawn to hypotenuse. Include paper millimeter rulers. Have readers measure all four triangles and make conjecture.

IV. Proof of conjecture from III.
  A. Refer to theorem "A right triangle can be circumscribed by a semicircle whose diameter is the hypotenuse of the right triangle."
  B. Cite the day's notes so students can look back to when theorem in IVA was covered in class notes.

V. Conclusion - statement of conjecture as a theorem
  A. Note formation of two isosceles triangles when median is drawn.
  B. Note of caution - theorem holds only for right triangles, and only for median drawn to the <u>hypotenuse</u>.

An outline of class notes to guide the progress of the Math Annotation Project.

The first draft of the project follows, along with some comments by the teacher/editor. Make sure you understand what each comment means. When you read the final version, note how the comments were incorporated into the Annotation Project.

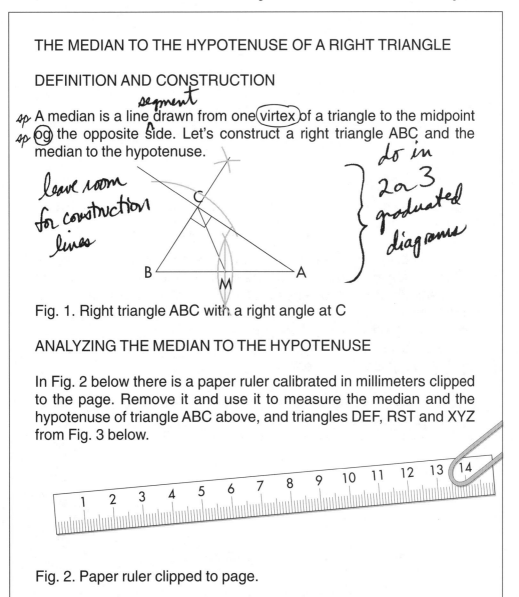

THE MEDIAN TO THE HYPOTENUSE OF A RIGHT TRIANGLE

DEFINITION AND CONSTRUCTION

segment

sp A median is a line drawn from one (virtex) of a triangle to the midpoint
sp (og) the opposite side. Let's construct a right triangle ABC and the median to the hypotenuse.

leave room for construction lines

do in 2 a 3 graduated diagrams

Fig. 1. Right triangle ABC with a right angle at C

ANALYZING THE MEDIAN TO THE HYPOTENUSE

In Fig. 2 below there is a paper ruler calibrated in millimeters clipped to the page. Remove it and use it to measure the median and the hypotenuse of triangle ABC above, and triangles DEF, RST and XYZ from Fig. 3 below.

Fig. 2. Paper ruler clipped to page.

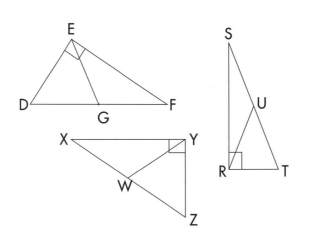

Fig. 3. Three right triangles, each with the median drawn to the hypotenuse.

give the measurements in a table.

Based on the found measurements, we conjecture that the measure
of the median to the hypotenuse of a right triangle is equal to one-half
the measure of the hypotenuse. We will call this the median-hypotenuse conjecture. Is the median-hypotenuse conjecture always true?

need new section heading here

The proof of the median-hypotenuse conjecture will require the use of
the following geometry theorem.

A right triangle can be circumscribed by a semicircle whose diameter
is the hypotenuse of the right triangle.
Let's explain this theorem using the diagram in Fig. 4.

refer students to where this was previously covered

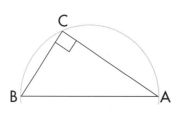

Fig. 4. The theorem states that A, B, and C lie on a circle with diameter AB.

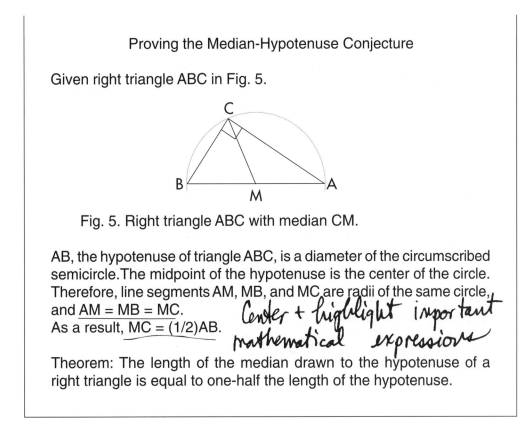

Proving the Median-Hypotenuse Conjecture

Given right triangle ABC in Fig. 5.

Fig. 5. Right triangle ABC with median CM.

AB, the hypotenuse of triangle ABC, is a diameter of the circumscribed semicircle.The midpoint of the hypotenuse is the center of the circle. Therefore, line segments AM, MB, and MC are radii of the same circle, and AM = MB = MC.
As a result, MC = (1/2)AB.

Center + highlight important mathematical expressions

Theorem: The length of the median drawn to the hypotenuse of a right triangle is equal to one-half the length of the hypotenuse.

First draft of Annotation Project with editor's comments.

The completed Math Annotation Project is shown next. Compare its coverage of the topic with the original notes. From which would a student who was absent better learn the material? Do you think doing the project gave the student author a better command of the material?

THE MEDIAN TO THE HYPOTENUSE OF A RIGHT TRIANGLE

DEFINITION AND CONSTRUCTION

A median is a line segment drawn from one vertex of a triangle to the midpoint of the opposite side. Let's construct a right triangle ABC and the median to the hypotenuse. First we construct the triangle. The steps are outlined below and the construction appears in Figure 1.

1. Draw line segment AB.

2. Draw ray AR.

3. Construct perpendicular from B to ray AR. Call intersection of perpendicular line and ray AR point C. Triangle ABC is a right triangle with its right angle at C.

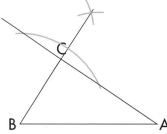

Fig. 1. Construction of right triangle ABC with a right angle at C

In Figure 2 below we construct the perpendicular bisector of the hypotenuse AB to find its midpoint.

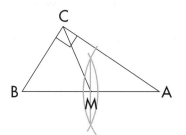

Fig. 2. Finding the midpoint M of hypotenuse AB

We then connect C to M forming line segment CM, which is the median, as shown in Figure 3.

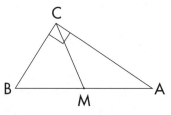

Fig. 3. CM is the median to hypotenuse AB.

ANALYZING THE MEDIAN TO THE HYPOTENUSE

In Fig. 4 below there is a paper ruler calibrated in millimeters clipped to the page. Remove it and use it to measure the median and the hypotenuse of triangle ABC above, and triangles DEF, RST and XYZ from Fig. 5 below.

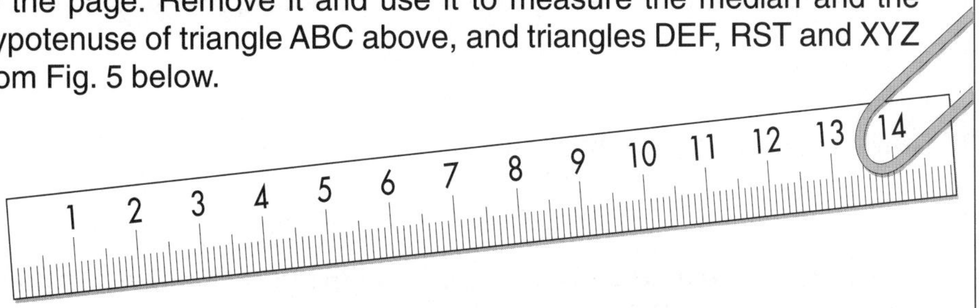

Fig. 4. Paper ruler clipped to page.

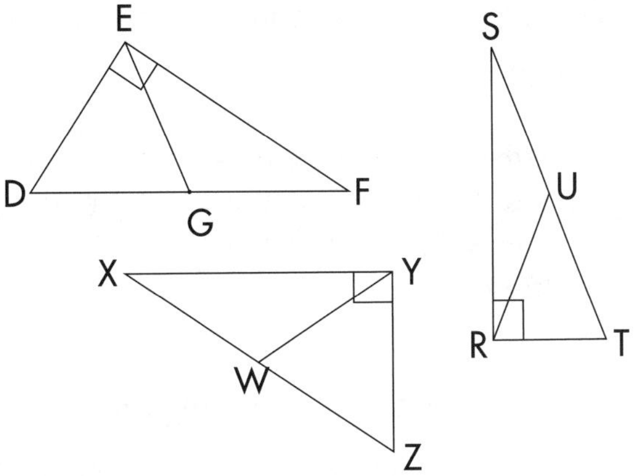

Fig. 5. Three right triangles, each with the median drawn to the hypotenuse.

The measurements are summarized in Table 1 below.

Table 1. Summary of Measurements for Four Triangles

TRIANGLE	LENGTH OF MEDIAN	LENGTH OF HYPOTENUSE
ABC	22 mm	44 mm
DEF	27 mm	54 mm
RST	24 mm	48 mm
XYZ	28 mm	56 mm

Based on the information in Table 1, we conjecture that the measure of the median to the hypotenuse of a right triangle is equal to one-half the measure of the hypotenuse. We will call this the median-hypotenuse conjecture. Is the median-hypotenuse conjecture always true?

PROVING OUR CONJECTURE

A Prerequisite Theorem

The proof of the median-hypotenuse conjecture will require the use of the following geometry theorem.

> A right triangle can be circumscribed by a semicircle whose diameter is the hypotenuse of the right triangle.

STOP & LOOK BACK!

> This theorem was covered in class during our unit on inscribed angles. A proof of the theorem and its converse can be found in the class notes from October 11.

Let's review this theorem using the diagram in Fig. 6.

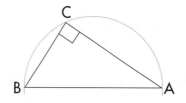

Fig. 6. Vertices A, B, and C lie on the semicircle with diameter AB.

The fact that the hypotenuse is the diameter will be crucial in our proof of the median-hypotenuse conjecture.

Proving the Median-Hypotenuse Conjecture

Given right triangle ABC in Fig. 7.

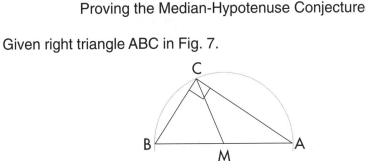

Fig. 7. Right triangle ABC with median CM.

AB, the hypotenuse of triangle ABC, is a diameter of the circumscribed circle.
The midpoint of the hypotenuse is the center of the circle. Therefore, line segments AM, MB, and MC are radii of the same circle, and

$$AM = MB = MC.$$

As a result,

$$MC = (1/2)AB.$$

Theorem: The length of the median drawn to the hypotenuse of a right triangle is equal to one-half the length of the hypotenuse.

Notice that when the median is drawn to the hypotenuse, two isosceles triangles are formed.

DON'T FROG-ET!!! This theorem holds only for **right** triangles and for the median drawn to the **hypotenuse**.

Sample completed Math Annotation Project.

The above Math Annotation Project enhanced the notes with the addition of constructions, a discovery activity, and explanatory notes. It is comprehensive and clear. Perhaps you would have added different enhancements in certain spots. Notice the use of the warning at the end. Use a dash of creative license in your Annotation Project. Enjoy it!

Evaluating Your Annotation Project

As you write your Annotation Project, following the steps given earlier, make corrections based on your proofreading and editing. Avoid submitting drafts with many errors by using the suggestions in this chapter as a checklist as you proofread and edit your work. Evaluate your project with respect to each of the following statements:

❏ The paper covers all the notes.
❏ The paper is mathematically correct.
❏ The paper is well-organized with respect to sections and paragraphs.
❏ The physical layout of the paper, including diagrams, tables, and word processing, is high-quality.
❏ Diagrams are graduated where necessary for clarity.
❏ The material is explained well (with correct sentence structure and accurate use of math terms).
❏ There are original examples as well as cited examples not from the class notes.
❏ The steps of proofs and derivations are adequately explained.
❏ The examples are appropriate, sufficient, and chosen with purpose.
❏ There are good explanations of typical pitfalls.
❏ Questions from handouts, tests, and quizzes are included and analyzed.
❏ There is a list of key terms correlated with page numbers.

❏ The use of technology (calculator keystroke sequences and computer programs) is explained.
❏ Technology is used with discretion.
❏ The list of writing tips (Chapter 3) was followed.
❏ The paper was submitted, edited, and revised in a timely fashion.
❏ All recommendations and corrections from edits are incorporated into the paper.

❏ The paper does a clearer job of explanation than the original notes.

❏ The general depth and quality are your best; you worked to your full potential.

Working on your Math Annotation Project will improve your understanding of mathematics and your writing skill. Additionally, the annotation projects done by other students can help you if you missed class or if something about a topic is unclear to you.

Compiling an Entire Year's Notes

You take notes in your math classes on a daily basis. The entire math course is usually divided into units that are two to three weeks long, and chapters in textbooks often form succinct units for this purpose. Your entire class may decide to do annotation projects. If at least one student is assigned to every unit, a compilation of the entire year's notes will be created. A whole-unit annotation project is a major assignment. Projects are completed and handed in throughout the year, not on one due date at year's end. This way, your editor (your teacher) works only with a few projects at any one time. Students writing manuscripts early in the year benefit from the work of other students in previous years. Students completing the project later in the year have an onus on them—not to repeat any errors that were previously cited.

The finished project for each unit should be placed in one large binder. It can be placed on reserve in the school library and made available to students who were absent or are studying for finals, for example. The complete volume of the entire year's work represents quite an accomplishment. This pseudo-textbook is a testament to what a class's cooperative effort can achieve when students strategically try to improve one facet of their mathematics—their ability to write effectively.

The Write Stuff

Employ your newfound writing skills when you take notes in class—add explanations and do more than just copy what is written on the chalkboard. Your ability to write mathematics will manifest itself in the quality of your research paper. You can do a Math Annotation Project concurrently with your math research. By the time you actually start writing your research paper, you will have improved your ability to communicate your findings.

Chapter Five
Conjectures, Theorems, and Proofs

Before you could speak, you were recognizing patterns and making generalizations. You could predict events of the day based on your experiences with the sequence of events in previous days. You even made conjectures before you could crawl! You hypothesized that it was near lunchtime, bedtime, and so on. You hypothesized that it was bath time when you heard the water running in the tub at a certain time of day. As a young adult, you make more sophisticated conjectures about everything from politics to fashion, and you have the ability to verbalize your conjectures. Did you ever realize how often you make conjectures in a day?

You are supposed to meet your friend Ryan in the cafeteria at 3:30 P.M. If by 3:55 he hasn't shown up, how many conjectures run through your mind? Your questioning and problem-solving skills are working so naturally you may not even realize they are operating!

- ❏ Did Ryan forget?
- ❏ Is this the wrong meeting place?
- ❏ What activity could have held up Ryan?
- ❏ Where should I look for him?
- ❏ Where was his last-period class?
- ❏ Was he taking the bus or walking home today?
- ❏ Might he be at his locker? Why or why not?
- ❏ Could my watch be wrong?

Making conjectures and testing are an integral part of mathematics research. Proof is the language by which much of mathematics is communicated. The ability to prove mathematical claims sets mathematics apart from other disciplines. With proofs, mathematicians can explain or demonstrate *why* their findings must be true. When someone makes a mathematical discovery and proves it, the discovery becomes incontestable.

The Role of Proof in Historical, Scientific, and Mathematical Research

Most students carry out their first research in a history class, usually by writing an expository paper on some aspect of history. Keep in mind that with historical research the reader can never be 100% certain that an argument used to back a theory is correct.

History papers can involve an original theory (the Kennedy assassination was planned by . . .), but a proof for such a theory is very hard to uncover because you are relying on the interpretation, selection, and validity of previously written sources. If, through convincing historical evidence, a theory based on historical research is accepted as "true," it would probably be difficult to say that it was *definitely* true, beyond any doubt. It would be hard to place a percentage value on whether or not the theory is true. Are you 99% sure? 60% sure? Is it reasonable to even try to apply a number to a qualitative argument? Probability is not a cornerstone of historical research, in part because historical research deals with events that have already happened and thus depends on outside sources.

The results of many scientific research experiments have a probability of being true. In certain scientific experiments, we can compute a probability that the results are accurate and base our conclusions on the rigor of the experiment and the value of the probability. You will learn in statistics that many scientific tests involving samples have a chance of being misleading based on the probability of obtaining an unrepresentative sample. The probability of obtaining accurate results can be computed, and it will never be 100%. If an experiment has a 99% chance of being correct, will scientists believe it is accurate? If the experimental design is solid, they will. Policies may be formed and further research will take place based on the results. If a scientist makes a hypothesis that is not supported by the results of an experiment, the result is still important. It may indicate that the hypothesis is *not* true.

While historians and scientists must explain why they believe their theories are probably true, the proof of a mathematical conjecture leaves no doubt of the conjecture's correctness as long as its proof is correct. A conjecture, when proved, becomes a theorem. When disproved with an argument or a counterexample, a conjecture is false. Until a proof (or disproof) is found, a math conjecture is much like a theory in a history research paper—we don't know if it is true. Even if we have a strong conviction that the conjecture is true, we can't explain why without a proof.

Unlike historical research, math research always involves the chance that a mathematician will discover a proof that we are 100% certain is correct. It may take centuries to finally prove something, and proof attempts—both successful and unsuccessful—can lead to whole new discoveries and the development of entire branches of study. Perhaps the most famous contemporary breakthrough in mathematics was the proof of what has been called "Fermat's Last Theorem." French mathematician Pierre de Fermat (1601–1665) conjectured that the equation

$$x^n + y^n = z^n$$

has no nonzero integer solutions for x, y, and z when $n > 2$. Fermat wrote, in the margin of a math text, "I have discovered a truly remarkable proof which this margin is too small to contain." Since no proof was discovered for over 350 years, it's generally believed that the proof Fermat had in mind was likely incorrect. In June 1993, British mathematician Andrew Wiles offered a proof, but Wiles withdrew his claim when he and other mathematicians identified problems with the proof later that year. By 1995, it seemed likely that Wiles had been able to correct his proof. Over the years, unsuccessful attempts to prove the theorem led to the discovery of commutative ring theory and a wealth of other mathematical findings.

Proofs form the cornerstone of much mathematics research. Before you can create proofs on your own, however, you need experience in reading and explaining the completed proofs of others.

Theorems and Proofs as Part of Your Research

Most of the articles you read will contain proofs of the claims that are made. Before you become actively engaged in reading mathematics, you need to become familiar with the types of proofs you will encounter in your readings. Perhaps you have some experience in writing proofs from units in geometry or logic studied in your math classes. At this point in your mathematics education, your experience with proofs might be limited. A good way to acquire knowledge about proofs is to read through proofs in journal articles or in a math textbook. See if you can explain the reasoning behind each line in a given proof. Reading and explaining "between the lines" will help orient you to the process of proofs. It can be especially helpful to do this with another student or in a small group. Exposure to finished proofs is a building block in learning to create original proofs of your own. In your research project, you can handle theorems and proofs on several different levels.

- ❏ You can explain the reasons for each step of the proofs given in the articles you cite. Often, the reasons are omitted because the target audience of the journal is mathematics teachers, who have experience with proofs.
- ❏ You can add steps in between the lines of the proofs given in an article. It is often assumed that the readership of the article can figure out the steps that are missing. In your research project, you can present and explain these missing steps.
- ❏ You can test the claims made in an article. If a theorem is presented, whether or not it is proved, try a few examples to verify the claim. Include these original examples as illustrations of the theorem in your research project.
- ❏ You might make a conjecture based on information found during the course of your research. Illustrate your conjecture with original examples. At times, you might disprove your own conjecture with one of your examples. The conjecture and counterexample can still be included in your research project.
- ❏ You might actually try to prove a conjecture you make. If so, you may need some guidance from your teacher in planning such proofs.

Tailor your treatment of proofs to your research. This chapter offers an overview of proofs similar to the overview of problem solving in Chapter 2. Just as Chapter 2 couldn't possibly provide all the experience you need in problem solving, the aim of this chapter is merely to touch on a few examples of proofs, primarily to help you in your reading. You'll need to draw on other math class experience to become an accomplished writer of proofs. When you read the chapter, concentrate on the direct proofs taken from actual student papers. Use this chapter as a reference as you encounter proofs in your article. Make sure you give the readers of your research thorough explanations of the proofs related to your article, whether they were written by you or by the article's author.

Making Mathematical Conjectures

Many of the conjectures you make in daily life require estimating and having number sense. Hypotheses about money, time, temperature, and so on are somewhat mathematical because they involve numbers. In your research, you will usually make conjectures based on purely mathematical situations. Let's take a look at some mathematical conjectures.

> Find the next two terms of the sequence shown:
> 2, 4, 6, _____, _____, . . .

For most people, the numbers 8 and 10 seem logical. Are they the only choices?
❑ If the rule is "Add the previous two numbers to find each term," then the next two numbers are 10 and 16.
❑ If the rule is "Start with 2 and 4, and add all previous numbers to find each term," then the next two numbers are 12 and 24.
❑ If the rule is "Start with 2 and 4, multiply the two previous numbers, and then subtract 2," then the next two numbers are 22 and 130.

Which answer is correct? Based on the information given, all are equally logical. There are probably dozens of other possibilities. But how many people are mathematically open-minded enough to use their imaginations and insight to search beyond the common pattern 2, 4, 6, 8, 10 ? Your ability to study patterns for their obvious and not-so-obvious properties is crucial to your ability to conjecture. Examine the following set of patterns, determine in which sequence the number 15 would fit, and explain the rule used to generate that pattern.

> 1, 4, 7, 11, 14, . . .
> 0, 3, 6, 8, 9, . . .
> 2, 5, 10, 12, 13, . . .

Be open-minded! If you don't want to see the solution, read no further. If you are working on arithmetic algorithms to find a logical connective between each of these terms, you are probably going to get frustrated. The number 15 would be the next term in the last sequence. Why? The first sequence is whole numbers written with straight-line segments only. The second sequence is whole numbers with curved digits only. The third sequence is whole numbers that are written with both curved and straight lines. The moral? Use your "mathimagination" in searching for patterns.

As your mathematics background, intuition, and insight develop, you will learn to make conjectures based on patterns you observe. Some conjectures may turn out to be true; others false. When you make a conjecture, it must be reasonable based on the information given. That is, the conjecture must hold for the cases you've seen. In Chapter 1, you encountered several patterns and conjectures as a preview. Now you are ready to concentrate on forming simple and then more complex mathematical conjectures.

Study each of the following examples and try to form conjectures based on the information given.

❑ The list of perfect squares is infinite: 1, 4, 9, 16, 25, 36, ... What conjectures can you make about patterns in this sequence?

❑ Inspect this list of Pythagorean triples:

      3, 4, 5

      5, 12, 13

      7, 24, 25

      9, 40, 41

      11, 60, 61

      13, 84, 85

What conjectures can you make about the patterns in this sequence?

❑ The set of prime numbers is infinite:

$\{2, 3, 5, 7, 11, 13, 17, 19, 23, 29, 31, 37, 41, \ldots\}$

What conjectures can you make about the prime numbers?

❑ The Fibonacci sequence is a recursive sequence (see Chapter 1):

$\{1, 1, 2, 3, 5, 8, 13, 21, 34, 55, \ldots\}$

Try making some conjectures about the Fibonacci sequence.

After considering each of the four sequences above independently, can you think of some conjectures based on combinations of the above concepts? Are there Pythagorean triples composed of all prime numbers? Of only composite numbers? Are any perfect squares prime numbers? Are any Fibonacci numbers perfect squares? Is there a number that is prime, a Fibonacci number, and the length of the hypotenuse in a Pythagorean triple? As you ask yourself probing questions based on the sequences and how they are formed, you might decide to test some more cases. Formulate a conjecture based on your question and its truth for the limited number of trials you made.

If you were doing research in the eighth grade on the following topics in geometry and number theory, do you think you would have formulated some of these conjectures?

❏ The base angles of an isosceles triangle cannot be right angles.

❏ The diagonals of an isosceles trapezoid are congruent.

❏ The sum of the lengths of any two sides of a triangle is greater than the length of the third side.

❏ The area of a square with side x is greater than the area of a circle with diameter x.

❏ If the sum of two prime numbers is odd, one of the numbers must be 2.

❏ If a parallelogram is not a rectangle, its four vertices cannot lie on one circle.

❏ The number 123! is not prime.

In the next section, we will look at some conjectures made by students and describe several types of proofs used in mathematics.

Examples of Mathematical Proofs

Before examining different mathematical conjectures and their proofs, we offer a warning necessary in the light of the tremendous technological advances being made today. As computers and graphics calculators become as common as paper and pencil, we are gaining a terrific tool for finding patterns and making conjectures. Machines can test many cases, allowing us to present quite a convincing argument for a conjecture that holds true after extensive testing. Keep in mind that these long lists of numbers that result from extensive testing do not constitute a proof. For example, a computer printout of the first 125,000 twin-prime pairs (see Chapter 1, page 6) does not mean there are infinitely many twin-prime pairs. Such lists certainly can display enough data to make our effort to *find* a proof a reasonable undertaking, but they are not proofs in and of themselves. We will discuss direct proofs, indirect proofs, and proofs that use the principle of mathematical induction.

Direct Proofs

There are times when conjectures can be proved by testing every possible case. This is not common, but it can occur, and it is one form of direct proof. In fact, it's one type of proof that a computer is capable of performing. Let's look at two examples:

❏ Julianne was reading an article from the February 1991 issue of *Mathematics and Informatics Quarterly*, titled "What Is the Use of the Last Digit?" The article was written by Lyubomir Lyubenov. The first part of Julianne's research required her to prove that, for all whole numbers x, the units digit of x^5 was the same as the units digit of x. Because the units digit is the only one in question here, Julianne raised each of the digits 0–9 to the fifth power and found that the units digit of x^5 is indeed equal to the units digit of x.

This result can be obtained rather quickly either by hand or by using a scientific calculator, and it constitutes the proof. Can you use this fact to prove that $x^5 - x$ must be divisible by ten? (Lyubenov, 1991.)

❑ Ben was doing research on the Platonic solids. There are five Platonic solids: the tetrahedron, cube, octahedron, dodecahedron, and icosahedron. Ben counted the number of faces, edges, and vertices on each Platonic solid. After working with just three of the solids, he formulated a conjecture. He tested his conjecture by counting the parts of the other two solids and found that his theory was correct. His proof was complete because he tested the data for all possible cases. Later he was able to find alternate proofs of his conjecture. Ben then wondered why there were only five Platonic solids and set out to prove that fact. He also planned to extend his work into other types of solids.

Perhaps you have used direct proofs in geometry. Let's review the proof of the following theorem: If two sides of a triangle are congruent, the angles opposite those sides are congruent. The figure below shows triangle *ABC* with angle bisector *BD*. Sides *AB* and *BC* are congruent. We want to show that angle *A* is congruent to angle *C*.

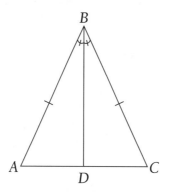

The plan is to prove triangle ABD congruent to triangle CBD. These two triangles already have two pair of congruent sides: *AB* is congruent to *BC,* and *BD* is common to both triangles. Angle *ABD* is congruent to angle *CBD* because *BD* is an angle bisector. Triangle *ABD* has two sides and an included angle congruent to two sides and an included angle of triangle *CBD*, respectively, so triangle *ABD* is congruent to triangle *CBD*. Since corresponding parts of congruent triangles are congruent, angle *A* is congruent to angle *C*. We have used deductive reasoning and theorems in geometry to prove that base angles of isosceles triangles are congruent.

You may have experience with proofs like this done in a two-column or a flowchart format, as shown on next page.

Statements	Reasons
1. $\overline{BD}$ bisects $\angle ABC$	1. Given
2. $\angle ABC \cong \angle CBD$	2. Def. of angle bisector
3. $\overline{AB} \cong \overline{CB}$	3. Given
4. $\overline{BD} \cong \overline{BD}$	4. Reflexive property of $\cong$
5. $\triangle ABD \cong \triangle CBD$	5. SAS
6. $\angle A \cong \angle C$	6. CPCTC

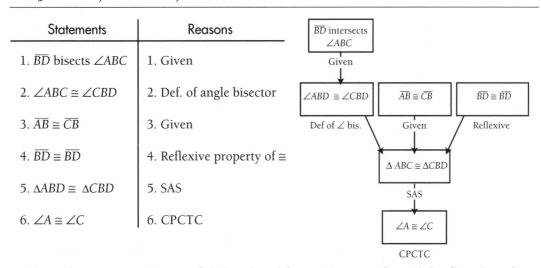

These formats provide a useful shorthand for writing proofs, and the flowchart form especially helps the reader visualize logical relationships. Paragraph proofs like that above, however, can offer commentary and greater detail. In the vast majority of math research, you will find proofs written in paragraph form.

Many direct proofs involve algebra. Variables are used because there are many, often infinitely many, cases to test. In Chapter 6, we will discuss criteria for picking a topic. For many high school students, the Pythagorean theorem is an appropriate introductory research topic. Let's look at a direct proof of the Pythagorean theorem itself, as well as some direct proofs from students' papers on Pythagorean triples.

Given a square, mark off equal sections of length a on each side of the square, as shown below. The remaining segments on each side have length b. The length of each side of the square can be represented by $a + b$. The congruent hypotenuses have length c.

The quadrilateral (rhombus) inscribed in the square is also a square. The four sides are equal because they are all essentially hypotenuses of congruent right triangles. Angles

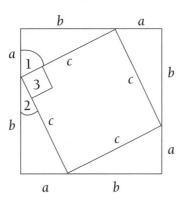

1 and 2 are complementary, since the acute angles of any right triangle are complementary. Since angles 1, 2, and 3 form a straight angle, angle 3 must be a right angle and the rhombus is actually a square.

The area of the larger square is $(a + b)^2$.

Squaring this binomial yields $a^2 + 2ab + b^2$.

The area of each of the four triangles is $(1/2)ab$.

The total area of the four triangles is $4(1/2)\,ab = 2ab$.

The area of the smaller square, c^2, can be expressed as the difference between the area of the large square and the total area of the four triangles:

$$a^2 + 2ab + b^2 - 2ab = c^2$$
$$\text{or } a^2 + b^2 = c^2.$$

Alexis did research on Pythagorean triples. Part of her work used the famous Euclidean method for generating Pythagorean triples. The Euclidean method involves formulas that require the user to choose two positive integers, p and q, where $p > q$, and substitute them to find the sides a, b, and c. (The derivation of these formulas could be part of a paper incorporating them.) The formulas are:

$$a = p^2 - q^2$$
$$b = 2pq$$
$$c = p^2 + q^2$$

Try some values of p and q and verify that Pythagorean triples are generated. Then,

Table 5.1 Some Pythagorean Triples Formed Using the Euclidean Method

p	q	a	b	c
2	1	3	4	5
3	2	5	12	13
4	3	7	24	25
5	4	9	40	41
6	5	11	60	61
7	6	13	84	85
8	7	15	112	113
9	8	17	144	145
10	9	19	180	181

prove that these formulas always satisfy the Pythagorean theorem. Experimentation with these formulas can lead to many interesting patterns and conjectures. Alexis used a spreadsheet to generate many sets of triples using this formula. Table 5.1 on the previous page shows a portion of her spreadsheet.

After fifty trials on the spreadsheet, Alexis noticed that whenever p and q were consecutive integers the larger leg and the hypotenuse differed by 1. She wanted to prove this conjecture; she believed it to be true. She started with the formulas that generate b and c. Then she substituted $q + 1$ for p in both formulas.

$$b = 2pq \qquad\qquad c = (q + 1)^2 + q^2$$
$$b = 2(q + 1)q \qquad\qquad c = q^2 + 2q + 1 + q^2$$
$$b = 2q^2 + 2q \qquad\qquad c = 2q^2 + 2q + 1$$

Alexis's direct proof shows that b and c differ by 1.

Sharon found another pattern using results from a table of Pythagorean triples generated using the Euclidean formulas (above) and a calculator. Looking at the triples generated using $q = 1$ and p varying from 2 through 12, Sharon noticed that the difference between the length of the short leg and the hypotenuse is always a perfect square. She conjectured that this was true for all values of p greater than 3, and set out to prove her theory.

$$q = 1$$
The short leg: $y = 2pq = 2p$
The hypotenuse: $z = p^2 + q^2 = p^2 + 1$
$$z - y = p^2 + 1 - 2p$$
$$= p^2 - 2p + 1$$
$$= (p - 1)^2$$
Since p is an integer, $(p - 1)^2$ is a perfect square.

Did you notice that $b = 2pq$ represented the shorter leg in Sharon's proof and the longer leg in Alexis's case? Can you explain why? Do you see how new discoveries lead to new questions and, ultimately, to new conjectures as we attempt to answer these questions?

Simran made a conjecture about primitive Pythagorean triples. **Primitive triples** are sets of three Pythagorean numbers that have the number 1 as their greatest common factor. For example, 3, 4, 5 is a primitive triple, but 6, 8, 10 is not. After generating many primitive Pythagorean triples, Simran noticed that the length of the hypotenuse was always prime or a multiple of 5. He made several attempts and "chipped away" at part of the proof that this was true for all such triples, but he could not complete the proof before school ended for the year. This became an excellent problem for future researchers to work on.

Laura was doing research on special properties of subtraction in different bases. She did many numerical examples to help her discover patterns. Some of her examples are shown here:

$$971 - 179 = 792$$
$$865 - 568 = 297$$
$$421 - 124 = 297$$
$$521 - 125 = 396$$

This part of her research inspired her to conjecture that if a three-digit number is written in ascending order of its digits and a new number is created using the same digits in descending order, the tens digit of their difference must be 9. Here is Laura's direct proof:

Given the number abc where $a > b > c$. From abc subtract cba; this will require regrouping since $a > c$.

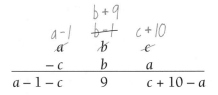

The patterns in the numerical examples led Laura to her conjecture. The proof showed that her conjecture was actually a theorem.

You've now seen several examples of direct proofs. Sometimes other types of proofs are more appropriate. Experience will help you decide what type of proof to use and how to execute your proofs. Let's take a look at another type of proof you might decide to use in your research—indirect proofs.

Indirect Proofs

Did you ever answer a question on a multiple-choice test by eliminating all the answers except one and then using that answer as your choice? If so, you did not *directly* answer the question. You found the answer *indirectly*, by showing that all the other possibilities could not be correct. This method is commonly used in mathematical proofs. Such proofs are called **indirect proofs**. If the other possibilities lead to statements we know are not true (creating a **contradiction**), then those possibilities do not hold. We start an indirect proof by making an assumption. This assumption is the only part of our proof that is in doubt; the rest of the proof follows all laws of mathematics. If we obtain a contradiction, then the only step that could be false is the original assumption. We can then switch our belief from the assumption to the other possibility. This process of proving theorems indirectly is sometimes called **reductio ad absurdum**. Let's take a look at two indirect mathematical proofs.

❏ The real numbers are the union of the set of rational numbers and the set of irrational numbers. Rational numbers can be expressed as the quotient of two integers; irrational numbers cannot. Prove that $\sqrt{2}$ is an irrational number.

There are two possibilities:

1. $\sqrt{2}$ is rational.
2. $\sqrt{2}$ is irrational.

Rather than proving directly that $\sqrt{2}$ is irrational, we will show that it can't be rational. Assume that $\sqrt{2}$ is rational. Then $\sqrt{2}$ can be expressed as the quotient of two integers, a and b, where a/b is in simplest form:

$$\sqrt{2} = \frac{a}{b}$$

Multiply both sides by b:

$$\sqrt{2}\, b = a$$

Square both sides:

$$2b^2 = a^2$$

The left side of the equation is even, since it has a factor of 2. Therefore, the right side of the equation, a^2, is even. If a^2 is even, then a must have a factor of 2 and thus is also even. If a is even, it can be represented as $2n$, where n is an integer. Then

$$a^2 = (2n)^2 = 4n^2.$$

Therefore,

$$2b^2 = 4n^2$$
$$b^2 = 2n^2$$

This tells us that b is even. Since a and b are both even, the fraction a/b is not in simplest form—it can be reduced by the factor of 2 in the numerator and denominator—contradicting the assumption that $\sqrt{2}$ can be expressed as a fraction in simplest form. Therefore, $\sqrt{2}$ cannot be expressed as a fraction in simplest form and must be irrational.

The following is an indirect proof from a student's paper on Pythagorean triples.

❏ Lauren used some mathematical notation in her proof of a conjecture about the **parity** (evenness or oddness) of numbers in Pythagorean triples. Lauren employed the symbol |, which means "divides evenly" or "is a factor of." (For example, 7 | 28 and 3 | 12.) She also used the arrow to represent "implies." Lauren conjectured that, in a primitive Pythagorean triple, only one of the numbers x, y, and z can be even. She proved this by showing that none of the other possibilities could work. Here are the four possibilities:

1. All three numbers, x, y, and z, are even.
2. Exactly two of the numbers are even.
3. Exactly one of the numbers is even.
4. None of the numbers are even.

Lauren ruled out choice 1 because if all three numbers are even, they have a common factor of 2 and do not form a primitive triple.

Lauren ruled out choice 4 by showing that if all the numbers are odd there is a contradiction. Assume x, y, and z are odd.

If x is odd, x^2 is odd.
If y is odd, y^2 is odd.
The sum of two odd numbers, x^2 and y^2, is even.

Therefore, since $x^2 + y^2 = z^2$, the number z^2 must be even, and since z itself is a whole number, z must be even. This contradicts our assumption that z is odd, so our assumption that all three numbers are odd is incorrect.

Lauren ruled out choice 2 in a similar fashion.

Assume x and y are even. This means $2 \mid x$ and $2 \mid y$. Therefore, for some integers m and n,

$$x = 2m \text{ and } y = 2n.$$

(These types of expressions are commonly used to represent even numbers in proofs. How could you represent odd numbers?) Substituting into the Pythagorean theorem:

$$(2m)^2 + (2n)^2 = z^2$$
$$4m^2 + 4n^2 = z^2$$
$$4(m^2 + n^2) = z^2$$
$$4 \mid \text{left side of equation} \rightarrow 4 \mid \text{right side} \rightarrow 4 \mid z^2 \rightarrow 2 \mid z$$

Since $2 \mid z$, z is even. Therefore, if two numbers in a Pythagorean triple are even, the third must be even. (Other proofs that start with the assumptions x and z are even or y and z are even follow similar steps.) Choice 2 is eliminated.

Indirectly, we have shown that the only possibility is choice 3: exactly one number is even. Interestingly, the hypotenuse z can never be even in a primitive Pythagorean triple. Given that exactly one of the numbers x, y, and z can be even, try to prove indirectly that z can't be even. (Hint: If a number is odd, it can be expressed as $2n + 1$, where n is an integer.)

If you want to gain more practice with indirect proofs, borrow a geometry, precalculus, or twelfth-grade math textbook from your school library or math department. Use the index to find some proofs and carefully read through them. See if you can re-create the proofs without looking at them. If you get stuck on proofs in your own research, you can work through them with your teacher during your consultations. It takes years to become adept at proofs in mathematics, so expect to proceed gradually. The more experience you get, the faster you will internalize the processes.

Next we will consider a sophisticated method of proving mathematical theorems usually first encountered by seniors in high school or by college students.

The Principle of Mathematical Induction

In your research, you may encounter infinite sequences of positive integers. You might make generalizations about patterns in these sequences after observing a finite part of them. You will naturally wonder whether your conjecture holds true for the entire sequence. Remember that extensive lists supporting your conjecture do not make it true. When you make a conjecture about the rule that determines a sequence, you are using **inductive reasoning**. You can test your generalization by trying more cases, but you could never try all the possible cases in an infinite sequence. Sometimes you can use algebra to prove such conjectures. The **principle of mathematical induction** is a process that can be used to prove conjectures about infinite sequences.

Before looking at some examples of proofs that use this principle, let's examine two common analogies to the principle of mathematical induction.

The Ladder—Picture a ladder with infinitely many rungs stretching into the heavens and outer space. If the following two conditions hold true, you can climb all the rungs:

    I. You are able to climb the first rung.

    II. Each rung is attainable from the rung directly before it.

If these two conditions hold, then you can climb the first rung and, by "chain reaction," you can climb each rung after it, indefinitely.

Dominoes—Imagine an infinite line of dominoes. This represents an infinite sequence. If the following two conditions hold true, you can knock down all the dominoes:

    I. You are able to knock down the first domino.

    II. Each domino is able to knock down the domino immediately following it.

If these two conditions hold, then you can knock down the first domino and, by "chain reaction," each domino will knock down the domino after it, indefinitely.

These two situations are analogous to using mathematical induction to prove conjectures about infinite sequences. First, show that your conjecture holds for the first number in the sequence. Then show that if your conjecture holds for the nth case, it also holds for the $(n + 1)$st case. If these two conditions hold, then your conjecture is true. Here is a formal statement of the principle of mathematical induction:

> Let $P(n)$ be a statement about any positive integer n. If
> I. the statement is true for $n = 1$ and
> II. whenever the statement is true for a positive integer k, it is also true for $k + 1$, then the statement $P(n)$ is true for all positive integers n.

We are now ready to demonstrate the principle of mathematical induction in action. As you study and read over each example, keep in mind the analogies of the dominoes and the ladder.

Table 5.2 shows the sum of the first n odd positive integers for several values of n. The term x_n represents the nth odd positive integer. Can you make a conjecture about the sum of the first n odd positive integers?

Table 5.2 Sums of the First n Odd Positive Integers

n	$x_n = 2n - 1$	Sum of Integers from 1 to x_n
1	1	1
2	3	4
3	5	9
4	7	16
5	9	25
6	11	36
7	13	49
8	15	64

It is logical to make the conjecture that the sum of the first n odd positive integers is n^2. Since this is a statement about an infinite sequence of positive integers n, we should try to prove it using mathematical induction.

> I. For $n = 1$: The "sum" of the first odd positive integer is 1, and $1^2 = 1$.
> II. Given that the sum of the first k odd positive integers is k^2, we must show that the sum of the first $(k + 1)$ odd positive integers is $(k + 1)^2$.

The kth odd integer can be represented by $2k - 1$. (Verify this with the numbers in Table 5.2.) Therefore, the $(k + 1)$st odd integer can be represented by $2(k + 1) - 1$.

The sum of the first $(k + 1)$ odd integers is equal to the sum of the first k odd integers and the $(k + 1)$st odd integer:

$$k^2 + 2(k + 1) - 1 = k^2 + 2k + 2 - 1$$
$$= k^2 + 2k + 1$$
$$= (k + 1)^2$$

Therefore, the sum of the first $(k + 1)$ odd integers is $(k + 1)^2$, if the sum of the first k odd integers is k^2. Since I and II hold, the conjecture is actually a theorem, proved by the principle of mathematical induction.

Table 5.3 shows values of

$$1/2 + 1/4 + 1/8 + 1/16 + \ldots + 1/2^n$$

and values of

$$1 - 1/2^n.$$

Table 5.3 Sums of Powers of 1/2

n	$1/2 + 1/4 + 1/8 + \ldots + 1/2^n$	$1 - 1/2^n$
1	1/2	1/2
2	3/4	3/4
3	7/8	7/8
4	15/16	15/16
5	31/32	31/32
6	63/64	63/64

You might make the conjecture that

$$1/2 + 1/4 + 1/8 + 1/16 + \ldots + 1/2^n = 1 - 1/2^n$$

We can try to prove this conjecture using mathematical induction.
  I. For $n = 1$, we have a true statement: $1/2^1 = 1 - 1/2^1$.
  II. Given that $1/2 + 1/4 + 1/8 + 1/16 + \ldots + 1/2^k = 1 - 1/2^k$, we must show that
    $1/2 + 1/4 + 1/8 + 1/16 + \ldots + 1/2^{k+1} = 1 - 1/2^{k+1}$.

We use the equation $1/2 + 1/4 + 1/8 + 1/16 + \ldots + 1/2^k = 1 - 1/2^k$ to make the substitutions in the parentheses below.

$$(1/2 + 1/4 + 1/8 + 1/16 + \ldots + 1/2^{k)} + 1/2^{k+1} = (1 - 1/2^k) + 1/2^{k+1}$$

Simplifying the right side of the above equation yields

$$1 - 2/2^{k+1} + 1/2^{k+1} = 1 - 1/2^{k+1}$$

We have shown that II holds. Since I and II hold, the conjecture is actually a theorem, proved by the principle of mathematical induction.

As you read each induction proof, go back to the two analogies to make sure you understand why conditions I and II cover all possible cases. If you need to apply the principle of mathematical induction in your research, first review this section. Then study the proofs given in precalculus and twelfth-year math textbooks. Read each proof and make sure you can explain each step. Then see if you can write them out without looking at the book. After you can do this, try the exercises in the textbook.

Disproofs and Counterexamples

Sometimes your findings will lead you to make a conjecture that you eventually find out is false. To disprove a conjecture, you need find only *one* case in which it doesn't hold. This case is called a **counterexample**. If you make a conjecture and find a counterexample, do not discount the importance of your work. Report it in your research paper. If it is reasonable to make such a conjecture, than the fact that you have disproved it is essential information for readers of your paper and for future researchers of your topic.

Let's examine some famous conjectures that were disproved along with some conjectures made by students and subsequently disproved.

❑ The famous mathematician Pierre de Fermat (1601–1655) thought that, for all whole numbers n,

$$2^{2^n} + 1$$

generates a prime number. This conjecture was based on the pattern shown in Table 5.4.

Table 5.4 The First Five "Fermat" Primes

n	$2^{2^n} + 1$	Prime?
0	3	yes
1	5	yes
2	17	yes
3	257	yes
4	66537	yes

When $n = 5$, the number $2^{2^n} + 1$ is equal to $4,294,967,297$.

It turns out that this number is not prime; it is equal to

$$641 \cdot 6,700,417$$

The lack of computers made finding factors of larger numbers almost impossible; you can see why the pattern inherent in the first five "Fermat" numbers could lead to the false conjecture.

❏ The ancient Chinese conjectured that if x is an integer greater than 1 and

$$x \mid (2^x - 2)$$

then x is a prime number (Stark, 1970). It turns out that, for $x = 341$, the conjecture does not hold.

❏ Alexis conjectured from a list of primitive Pythagorean triples that if the short leg is an odd integer, then the lengths of the longer leg and the hypotenuse are determined. In other words, there are not two primitive Pythagorean triples with the same odd-numbered short-leg length. Her conjecture was based on many trials. Eventually, Alexis found the following counterexample: The triples

$$33, 56, 65 \text{ and } 33, 544, 545$$

are both primitive Pythagorean triples with the short leg equal to 33. A new question arises. Are there other Pythagorean triples with short-leg length equal to 33? Are there infinitely many more? Some experimentation will precede the formulation of a conjecture based on this question.

❏ Robin was researching the relationship between an equilateral triangle and the three largest possible nonintersecting circles inside it. Her work led her to an inequality involving six real numbers. At one point in her work, she conjectured that

$$\text{If } a + b + c < d + e + f, \text{ then } a^2 + b^2 + c^2 < d^2 + e^2 + f$$

where a, b, c, d, e, and f are real numbers. It seemed reasonable that the smaller sum would yield a smaller sum of squares. Try some numbers and see if you agree. Robin found a counterexample:

$$a = 0.2$$
$$b = 0.2$$
$$c = 3$$
$$d = 0.5$$
$$e = 1$$
$$f = 2$$

Verify that Robin did, indeed, find a counterexample. Can you look at the numbers she used and determine her strategy in finding the numbers? Try to find a counterexample of your own.

Undetermined Conjectures

Any conjecture you make is either true or false. You may not know the truth value of some of your conjectures, but a truth value exists. We call such conjectures **undetermined conjectures** because their truth value is, as of now, unknown. You won't be able to prove or disprove every conjecture you make. Your conjecture will still be valid, and another researcher may attempt a proof or counterexample based on your work. That is why it is essential to offer undetermined conjectures in your research. You should always try to prove or disprove the conjectures you formulate, but when you can't, for whatever reason, do not shy away from reporting them.

Jocelyn did research on Pythagorean triples and Fibonacci numbers based on the article "Pythagoras Meets Fibonacci" in the April 1989 issue of *Mathematics Teacher*. The article mentioned some relationships between the Fibonacci numbers, the sides of the associated right triangles, and the areas of these triangles. Jocelyn wondered about the relationships between the Fibonacci numbers, the associated right triangles, and the *perimeters* of the triangles. She created a spreadsheet, shown in Table 5.5 on the next page.

Although you may not have read the four articles Jocelyn read, you can study the table and look for patterns in it. Jocelyn finished her paper by offering several conjectures for other researchers to pursue:

- ❑ All of the perimeters are even numbers.
- ❑ Every other consecutive pair of perimeters is evenly divisible by 3.
- ❑ Beginning with 30 and 80, there are three perimeters not divisible by 10 followed by a pair of perimeters that are evenly divisible by 10.
- ❑ The perimeter is twice the product of the third and fourth Fibonacci numbers.
- ❑ The four Fibonacci numbers all divide evenly into the associated perimeter.
- ❑ The perimeter equals the sum of the squares of the third and fourth Fibonacci numbers, minus the square of the second Fibonacci number.
- ❑ If the perimeter is evenly divisible by 3, then the product of the first two Fibonacci numbers is not evenly divisible by 3.
- ❑ If the perimeter is not evenly divisible by 3, then the product of the third and fourth Fibonacci numbers is divisible by 3.
- ❑ The perimeter on the nth line is equal to the middle leg on the $(n + 1)$st line.

Jocelyn's conjectures could actually spawn an entire research paper for a reader of her work. Formulating them shows the insight and expertise that come with effort and practice.

Table 5.5 Jocelyn's Spreadsheet

Four Consecutive Fibonacci Numbers $a, b, a+b, a+2b$	Pythagorean Triple $a^2 + 2ab, 2ab + 2b^2, (a+b)^2 + b^2$	Perimeter $x + y + z$
1, 1, 2, 3	3, 4, 5	12
1, 2, 3, 5	5, 12, 13	30
2, 3, 5, 8	16, 30, 34	80
3, 5, 8, 13	39, 80, 89	208
5, 8, 13, 21	105, 208, 233	546
8, 13, 21, 34	272, 546, 610	1428
13, 21, 34, 55	715, 1428, 1597	3740
21, 34, 55, 89	1869, 3740, 4181	9790
34, 55, 89, 144	4896, 9790, 10946	25632
55, 89, 144, 233	12815, 25632, 28657	67104
89, 144, 233, 377	33553, 67104, 75025	175682
144, 233, 377, 610	87840, 175682, 196418	459940
233, 377, 610, 987	229971, 459940, 514229	1204140
377, 610, 987, 1597	602069, 1204140, 1346269	3152478
610, 987, 1597, 2584	1576240, 3152478, 3524578	8253296
987, 1597, 2584, 4181	4126647, 8253296, 9227465	21607408
1597, 2584, 4181, 6765	10803705, 21607408, 24157817	56568930
2584, 4181, 6765, 10946	28284464, 56568930, 63245986	148099380
4181, 6765, 10946, 17711	74049691, 148099380, 165580141	387729212
6765, 10946, 17711, 28657	193864605, 387729212, 433494437	1015088254
10946, 17711, 28657, 46368	507544128, 1015088254, 1134903170	2657535552
17711, 28657, 46368, 75025	1328767775, 2657535552, 2971215073	6957518400
28657, 46368, 75025, 121393	3478759201, 6957518400, 7778742049	18215019650

Proving and Improving

Becoming adept at mathematical proofs takes time, patience, and effort. At several junctures in this chapter, readings were suggested to help you further your experience with mathematical proofs. If you have questions about proofs, looking up the following words in the index of a book that includes mathematical proofs should help you:

- ❏ conjectures
- ❏ counterexample
- ❏ deductive proofs
- ❏ deductive reasoning
- ❏ direct proofs
- ❏ disproof
- ❏ indirect proofs
- ❏ inductive reasoning
- ❏ principle of mathematical induction
- ❏ reductio ad absurdum

Read the sections indicated in the index for each word. Try the model problems that offer solutions. See if you can give the reasons for each step in a completed proof. Annotate the proofs. Your confidence and agility with proofs will develop with practice and exposure to proofs. Keep a reference list of books that you find to be good sources of examples of proofs. If, during your research, you have problems with proofs, it will be helpful to know exactly what source you can go to for assistance. Discuss all original proofs and attempted proofs with your teacher. The degree to which proofs play a role in your research will depend on the topic you choose; Chapter 6 will help you with this step.

Chapter Six
Finding a Topic

Chapters 1–5 discussed the role of questioning, problem solving, writing, and proofs in mathematics. Chapters 6–9 will direct you through your research project. You have probably written many reports during your school career, but you may never have done a research project. How does a research paper differ from a report?

Reports and Research Papers

When a teacher assigns a report, most often it involves your going to a library. Reference materials from a library or computer service often provide the information necessary to construct a report. The material you gather and read was probably written by experts in the field. A report on the Civil War can include any information on the Civil War. You select the material you will include, organize it, write an outline, and write your paper. You proofread and revise your paper several times to make it as professional as possible. Writing a report is often a good way to orient yourself to a topic you are unfamiliar with. Reading a high school student's paper on the Civil War might be a good way for a tourist from another country to get an encapsulated introduction to the Civil War. As the author of a report, you are a *reporter*.

As a *researcher,* you start with a problem—a question that needs to be addressed. You read related material to try to formulate an answer to the question. You may have to do an experiment or compile evidence to support the solution you present. Original material can be added where necessary to help solve the problem. You generate other questions based on your readings and the "tinkering" you've done with the problem. Only information that contributes to the solution of the problem is included in the main body of the research paper. Material tangential to the narrow focus of the research problem is not included in the main body. Some of these "tangents" might be recommendations of topics for other researchers to study. You need to stay focused on explaining aspects of your central problem. You become a specialist on your problem. You are a participant in the creation of material for the research paper. At times, you may get stuck and need assistance in attacking a part of your solution. You might consult a mentor, an expert in the field, or your library, or you may put out a memo on a computer bulletin board. Throughout your research project, the central focus is the solution of a specific question or set of questions.

Focusing on one problem tends to keep research papers very specific. A student could write a math report on the Fibonacci sequence, but this would be too broad for a math research paper. "The Use of Fibonacci Numbers to Create Pythagorean Triples" is a specific topic that could be the focus of a math research paper. Topics such as "Probability,"

"Geometry," or "The Pythagorean Theorem" are too broad. The following are examples of topics that have a specific focus and would make suitable research paper topics:

❏ determining the conditions necessary for a quadrilateral to be able to be both inscribed in a circle and circumscribed around a circle

❏ the ratio of the radii of the inscribed circle to the circumscribed circle of a right triangle with integer-length sides

❏ using probability to find the area of irregular figures

❏ using the area of a circle to find the area of an ellipse

As you can see, the title of a research paper has the potential to be long because it must be descriptive. Chapter 8 discusses titles and other components of the formal paper in detail. Your main concern here is understanding that the topic you pick will have a narrow scope. As a newcomer to math research, you can use journal articles to provide the problem for your paper. Each article solves a specific problem and offers information essential to the solution of that problem. Using a journal article will help you pick a topic that is focused. Other factors that should guide your choice of a topic are discussed in the next section.

Choosing Your Research Project

Finding a topic is the first major milestone of your research project. If you have the option of choosing your own topic, you have a responsibility to yourself. You will be spending extensive time working with the topic, so you want to pick a stimulating topic that is appropriate for your background. How do you find a topic for your math research paper? When you have a choice as to what research project to tackle, remember your investment in the project and consider the three "Ex"s as you make your choice:

❏ **Excitement:** The topic should interest you. The topic you choose will often be an outgrowth of a topic you enjoyed in your regular math class. Sincere interest in your work will give you a real sense of accomplishment and satisfaction, and this usually manifests itself in the quality of the research and in the paper itself.

❏ **Experience:** You should be familiar with the topic and its prerequisites or with a closely related topic. Pick a topic that is an extension of material you've seen before. You will have more mathematical intuition for a topic you already know something about. Posing questions, testing and forming conjectures, and reading about extensions of the topic rely on a knowledge base.

❏ **Expertise:** You should have mastered some of the fundamental prerequisites for researching the topic. Feeling strong and confident is always helpful, and success breeds strength and confidence. For example, if your topic relies heavily on algebra and you have always excelled at algebra, you already have expertise in some aspect of your research.

Researchers who choose their own topics naturally pick extensions of work they are entrenched in. For example, a nutritionist doing research would naturally be interested in the effects of cholesterol on the human heart. An aircraft designer would naturally be interested in researching the advantages of a new wing design on commercial air flights. Researchers can be fanatical about their work—often, dedication and obsession are only by-products of a long-term commitment to their field. As a newcomer to research, how do you fit in?

Your Mathematical Experience

Because much of the territory you chart with your research will initially be foreign to you, familiarity with a topic will be important in your choice of a research topic. Examine this scenario:

> You walk the halls of your school every day. A boy you don't know passes you in the hall every day. You don't say hello to each other; you barely exchange glances. You don't know each other's names. Years go by, and this relationship stays the same. One summer, you go on a vacation to Paris. You get off the plane, and standing there in the airport in Paris is the same boy from the halls in school! Do you walk up and say "Hello" *now*? You bet!

In foreign surroundings, we all take comfort in a bit of familiarity. What does this have to do with your research topic? In the same way that you find comfort in a familiar face in unfamiliar surroundings, you will use mathematics you are familiar with as an anchor as you proceed through your research. As you learn new concepts, they will then become familiar and become additional anchors.

As a high school student, you have sat in math classes for over a decade, and the best is yet to come. The high school curriculum is packed with many thought-provoking, challenging units. Do you have any general preferences regarding mathematics at this time?

❏ Would you prefer researching pure mathematics or applied mathematics?
❏ Are you a whiz at algebra?
❏ Do you enjoy geometry because of its visual nature?
❏ Do you wonder how different types of graphs are related to their equations?
❏ Are you fascinated by number theory?

Perhaps you have already wondered about some specific mathematical ideas that you have encountered in school or in your daily life.

❏ Do you wonder how satellite "dish" antennas are engineered?

❏ What do you know about the Pythagorean theorem besides $a^2 + b^2 = c^2$?
❏ How is the graph of a parabola related to the quadratic formula?
❏ Can you imagine the Fibonacci sequence having some complex applications?
❏ How is trigonometry used to find distances in our solar system?
❏ How can graphs be used to find how a corporation can maximize its profit?

The possibilities are endless. As we mentioned previously, research topics must be focused. Picking a topic that is too broad is a common mistake students make when they are writing a mathematics paper without sufficient guidance. Students often pick a broad topic and find many library books on the topic, as if they were writing a report. They then try to cull different pieces of information from the separate books and come up with a gelled report. Having many sources is not necessarily an advantage in writing a math research paper. A lengthy bibliography does not make the paper stronger. It is actually very difficult to select and integrate material from several very comprehensive books and then write a paper. It is easier to internalize work that came from *you*.

Classically, people start researching because one specific problem is gnawing at them, not because they got the urge to write a synopsis of several thousand pages of others' work. A mathematician could spend an entire lifetime researching one particular problem in probability; writing a single paper on probability would be trying to cover too much. Your paper will not be a summary of the material in many books. Quite the opposite—it may be built out of only a few pages and your ability to inquire about, process, and explain mathematical thoughts. Given the role of mathematical experience, how can you find a focused topic if your mathematics background is somewhat limited? The following sections give you some suggestions.

Building On Short Articles

The best way to find a focused topic is to use an article from a mathematics book, a mathematics journal, or a math education journal. Appendix A lists several books of short articles that can form the basis of a research paper, as well as several popular journals. Additionally, your state's mathematics teachers association may publish a journal; see your math teacher or the department chairperson for this information. The journals are kept in the periodicals section of the library. Older copies are available on microfilm or microfiche; more recent issues may be found bound together by year or as separate publications. Your high school math department may have a collection of back issues of certain journals. A mathematics teacher in your school may have a collection of photocopied journal articles for you to look at. Check your school library, public library, and local college libraries. Plan to spend some time looking through the journals once you've found them.

In each issue, the journals feature articles less than ten pages long (usually three to six pages) on a specific concept in pure or applied mathematics. Often they describe some novel aspect of a common topic or provide a new twist to an old topic in a succinct yet comprehensive way. Thus, starting with a journal article provides the focus you need to write a successful paper. It is likely that the bibliography for your paper will consist of one item—the article that sets you in motion. That doesn't mean your paper will be simply a rehashing of someone else's work. Chapter 7 will explain why and how a six-page article can provide you with months' worth of work. Your objective in Chapter 6 is to find a topic for your research project.

Take inventory of the mathematics that interests you. There will be only a few articles in each issue of a journal. Look at the tables of contents. Plan to look over all the articles; try not to rule out articles solely on the basis of their title. Carefully read the first paragraph or two of articles that look interesting, and then skim through the rest of them. Don't expect to understand everything. Try to determine which articles are reasonable extensions of your current knowledge. Pick your favorite three articles in order of preference. Show the articles to your teacher. Your teacher can help you determine a suitable choice given your math background. Don't attempt something that is loaded with prerequisites with which you are unfamiliar. Follow your teacher's advice. Once you've chosen an article upon which to base your research, you may need to find other sources to support your research. You can also ask your teacher for textbooks you can use to learn the prerequisite skills the article assumes the reader has.

These actual accounts of students' research will give you ideas for how students learned the prerequisite skills they needed in order to read and understand their articles.

❑ As a tenth grader, Daniele was working with probability and the quadratic equation. She needed experience with conic sections, specifically, the parabola. Daniele's first research task was mastering the graphs, equations, and properties of parabolas.

❑ Jeff's research on special cases of similar polygons required a knowledge of proportions in the right triangle. An old geometry textbook provided all the information Jeff needed.

❑ Simran's original proofs of properties of skewsquares required familiarity with proportions in similar triangles. These proportions were explained in an integrated mathematics textbook.

❑ Midway through Luke's research on the areas of irregular plane figures, he needed to learn about the trapezoidal rule, a topic from calculus. Luke looked in the index

of a calculus book, found the two-page treatment of the trapezoidal rule, and studied it before continuing with his research.

❏ Katie's work on number theory required her to learn about numbers in bases other
than ten. The article was heavily dependent on mastery of this concept. The information on other bases can be found in a middle-school textbook.

❏ Part of Robin's work on a famous problem involving triangles and circles required
that she learn partial derivatives and optimization theory. This is an example of an
article with too many difficult prerequisites, and as a result this part of the article
was not researched. The first part of the article provided her with excellent material on which to build a paper.

If you set aside a few hours, you can look through dozens of articles. Hopefully, the
suggestions for finding sources and articles given above will get you started. Keep in
mind that you may not be able to finish researching your article in the school year. On
the other hand, you might have time to read several articles on related topics and combine the results in your paper. Also remember that you need to pick an article appropriate for your background. For example, a high school student interested in the effects of
high blood pressure on sleep disorders in children cannot expect to do a research paper
on that topic that is as comprehensive as one written by a doctor who has specialized in
sleep disorders for the past twenty years.

After you have chosen the article you plan to use, make two or three photocopies of it.
You will want to work with photocopies, because you do not want to mark up an original
copy of an article. You will need an archive copy, which stays at home in case your other
copies are lost, and an annotation copy, which you work on. You might want to keep an
extra copy in your locker in case your annotation copy is misplaced. You may decide to
read several related articles and combine some of the findings to create a new result.
Make copies of all the articles you intend to use.

Other Sources for Research Topics

Journal articles are most highly recommended as sources for your research topic because
they provide direction in a succinct, comprehensive way well-suited to new researchers.
In this section, we discuss other sources for a topic. Keep in mind that these sources will
not provide as much written material as journal articles; consequently, from the outset,
you will be doing lots of probing on your own. Here are some suggestions:

1. **Using a problem as a springboard:** In your problem-solving work, you may encounter a problem that is only one to four sentences long that really intrigues you. After
you find a solution, ideas may fill your head as to how the problem could be altered
slightly, extended, or changed radically to provide new challenges. The new problems
you pose and their solutions and/or proofs could comprise a research paper totally in-

spired by that one problem. Read the following problem:

Rich has a devised a carnival game. He has a large square board tessellated with 2500 one-inch squares, in a 50-by-50 pattern. The squares are red or black and are in a checkerboard pattern. A person who wants to play the game tosses a dime on the board. If the dime lands totally within a red square, the player wins. If it lands in a black square or overlaps several squares, the player loses the dime. What is the player's probability of winning?

A common solution involves solving a simpler problem. Think of all the questions that might arise.

- ❏ What is the diameter of a dime?
- ❏ What if there were only 100 squares?
- ❏ What if the game used quarters?
- ❏ What if the board used triangles? Hexagons?
- ❏ Which polygons do and don't tessellate? Why?
- ❏ What if all of the squares were red, with black lines dividing them into 2500 squares?
- ❏ What if the board contained infinitely many squares?
- ❏ Could you pretend that the board has just one square and find the solution to that problem, and then adjust it for the 2500 squares?
- ❏ How would the theoretical probability change if the "board" were three-dimensional and the player threw a sphere instead of a circle?

The original problem (appropriately footnoted), its solution, and your conjectures, extensions, and so on can make a great paper. You must have access to someone who can check the accuracy of your work frequently. Appendix A lists many problem-solving books, and such books are available in most bookstores and libraries.

2. Thinking of a research problem on your own: As a person attuned to mathematics, it is possible that a question might hit you and you decide to research it. Blake, an eleventh-grader, had such an experience. He wondered how he would find the area of the shaded intersection (he named it a "bicusp") of two circles situated as shown in the figure below.

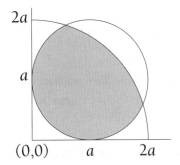

Blake's original conjecture involved an erroneous assumption about proportions, but finding the area proved to be a terrific exercise in analytic geometry, trigonometry, and calculus. The entire research paper was based on a query Blake had and on no other written inspiration. He did, however, use textbooks to learn concepts he needed in order to continue his work.

3. **Extending an idea from your core mathematics course:** As you engage daily in a core mathematics course, you have a chance to probe, conjecture on, and wonder about variations on what you are learning. You may decide to research an extension of a certain topic from your math class. For example, Laura studied about quadratic functions in her regular math class. She also had a brief introduction to simple absolute-value functions. Laura wondered about the graphs of quadratic functions that contain absolute-value expressions, such as the following:

$$y = |x^2 - 5x - 6|$$
$$y = |x^2 - 5x - 6| + 12$$
$$y = |x^2 - 5x - 6| - 4x$$
$$y = |x^2 - 5x - 6| + 3x^2 - 4x + 12$$

Laura's research on these graphs involved a computer, a graphics calculator, and pencil and paper. Her entire exploration was derived from topics covered in her regular math class. You might look for questions that could serve as research topics as you work on the Math Annotation Project described in Chapter 4.

4. **Using an application from real life:** You might be curious about a mathematically influenced aspect of the real world. Students have wondered about the shape of airplane wings, how skyscrapers are engineered, why honeycombs are based on the hexagon, how codes were used in wartime, how suspension bridges are built, and how interest is compounded, to name just a few topics. Mathematics is used in music, medicine, sports, astronomy, and many other fields, and these fields have inspired many math research papers. For example, Gino had often wondered why car radio antennas had a practical shape, a telescoping mast, while satellite "dish" antennas seemed so large and clumsy. He wondered about the shape of the satellite antenna. His research led him to journal articles about these antennas. He also needed some material from a textbook and from his math course, because the antenna's shape is based on two conic sections, the parabola and the hyperbola. In addition to reading his articles, Gino contacted companies that manufactured and/or used the large antennas.

Excellent application problems abound in mathematics. Applications can make good research topics, but be forewarned that you can easily be led into other disciplines, such as science or engineering, that may require years of background for you to fully understand a particular research topic. On your own, you are more likely to be able to find new patterns in and experiment with the Pythagorean theorem, the Fibonacci sequence, and so on, than to improve upon the design of the Golden Gate Bridge. Keep this in mind as you choose your topic.

Getting Started

You have been introduced to several ways of finding a topic for your math research paper. Allocate a reasonable amount of time to find out where you can locate the materials described in Appendix A. Keep in mind that the articles provide more focus than the other sources mentioned in this chapter. Get your copies and start on your topic as soon as you finish reading this chapter. You will be working on your research and reading *Writing Math Research Papers* concurrently. Refer to previous chapters in the book, the time line in the Introduction, and your teacher if you need assistance.

Try to stick with a topic once you have found one. If you change topics, the time you have invested in the original topic will be lost, and you will have less time left in the school year to complete work on another topic. That's why it is best to use care, discretion, and teacher guidance when you are initially choosing your topic. It is also wise to make your copies and begin your reading immediately. As you do your research, you will generate your own questions and try to answer them. You will be actively engaged in the research. Because you will not formally write your research paper until your research is well under way or, possibly, close to being finished, you will keep a journal of notes about your readings and findings. Chapter 7 will assist you in this next stage.

Chapter Seven

Reading and Keeping a Research Journal

You may never have thought about the different ways you read. You may read a novel quickly. You might read the newspaper comics with a radio playing. You may just skim parts of the newspaper's sports section looking for the results of your favorite team's game. You might skip whole sections of magazine articles as you search for a part that interests you. Road signs can be read at 55 mph, with a passing glance. You probably couldn't read an automobile insurance policy without jotting down questions. You would read a cake recipe or instructions for building a picnic table with care; an airplane-repair manual would be read slowly and scrutinized even more carefully. Reading mathematics materials from your textbook or a journal article has some idiosyncrasies of its own.

Preparation for Reading

When you read mathematics articles, allow ample time for each reading session. You might be able to read a newspaper article during a five-minute wait at the dentist's office, but you shouldn't allocate time for your mathematics reading in such short capsules. You'll need time to read, think, and digest information, and you'll always be writing and highlighting when you read mathematics. The degree of concentration necessary to interact with your reading precludes distractions, so don't leave the television or radio on as you read. As mentioned in Chapter 3, you'll probably notice the difference between the physical layout of a mathematics article and that of a novel. The pages of the mathematics article are broken up, not filled with continuous text. In a mathematics article, portions are centered and highlighted on their own lines. Often these portions are mathematical expressions. In most cases, they are a signal to the reader to stop and digest the information. Keep this in mind when you read and when you write mathematics.

Setting Up a Bibliography Tree

Each article you read will have a listing of references used by its author. It is helpful to make copies of some of these referenced articles in addition to your original article. They will add to your research, enhance your background, and possibly lead to other avenues of research. You don't need to read them initially, and you may never need them as reference, but having access to them could make your research stronger. They may offer key theorems, proofs, examples, questions, and hypotheses. Each article listed in your reference section will also have its own reference list. Construct a bibliography tree that represents a hierarchy of the development of your topic as represented by your articles. Nicole's bibliography tree for her research on properties of Pythagorean triples is shown here.

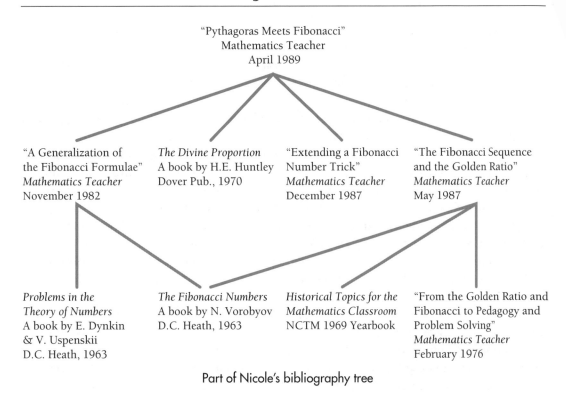

Part of Nicole's bibliography tree

Bibliography trees can become very lengthy; use discretion as you construct your tree. Some of the articles may not be helpful. Articles in popular journals may be easier to find than books that are many years old. Certain books contain dozens of references while some articles have no references. You can't read every book and article in the limited time you have, but certain articles that directly relate to your research can be beneficial. As you delve deeper into your topic, you may need one of these other sources, and you'll be glad you took the time to set up the tree and copy some of the supporting articles. You can start reading your main article and doing your research without having the other articles in hand—just keep in mind that you might want to access the other articles at some point in the course of your research.

Beginning a Research Journal

The formal research paper will not be started until much reading and journal writing have taken place. The journal houses the "guts" of your research. It will be your guide to transforming your research into a research paper. The journal is a binder filled with lined paper, blank paper (for diagrams), and graph paper (for charts and graphs) as you begin your reading. A binder is recommended because it will often become necessary to re-

e pages, hand them in for checking, rearrange them, rewrite them, and so on. A
.nder can also house any other papers that can be three-hole-punched, such as computer printouts and copies of articles. For these reasons, a spiral or bound notebook is
not practical for use as a research journal.

Perhaps you have kept journals in other classes; possibly even in other math classes.
Often these journals are reflective—they record your reactions to and feelings about
work that you've done. In such journals, being introspective may help you improve your
skills because you are forced to analyze not only your mistakes but why you think you
made them. This sensitivity can be tapped in future work. You'll remember how you felt
about your performance because the act of writing it down clarified it.

Your research journal, however, will not be a reflective piece. It will consist of the
written work you create as you read your articles and take notes, test claims, make conjectures, try proofs, and so on. The journal must be organized. Pages of notes in the
journal should be marked to correlate with the short notes on your annotation copy.
Make sure you date the top of all your notes. Much of your journal writing will look like
scratch notes, unpolished mathematical material you used to accentuate your readings.
But be careful—you must be able to decipher these notes when you incorporate them
into your final research paper. The meanings of short notes and abbreviations that you
understood at the time your notes were written may be forgotten when you go through
them months from now. Rewrite parts of your journal as you complete them so they are
legible and thorough. Keep the notes in the logical order that follows the progression of
your research. You can write up each section of your research as you finish it, using the
writing tips in Chapter 3. Then it will be easier to merge them into your formal paper. A
well-kept journal will translate efficiently into a research paper when the formal writing
stage begins.

The first pages of your journal should be a time line of your progress. The dates on the
time line will help you correlate the journal material with the parts of the articles they
refer to. Before continuing, take a look at Sara's research journal progress calendar (next
page), which shows her work schedule for her paper on escribed circles.

Apportioning the Readings

It is important to set up a research schedule and adhere to it. You should meet with a
teacher or mentor periodically during the course of your research project. Perhaps you
already meet with your teacher for extra help when you have questions about homework
or class work. The extra help sessions for your paper will be called consultations. The
project should proceed gradually. Meeting with your teacher provides time to ask questions, discuss the notes you take in your journal, and plan for the next reading. The
consultations may take place as needed, but they should occur at least once every two
weeks to keep the project progressing consistently. The amount of time spent at a consul-

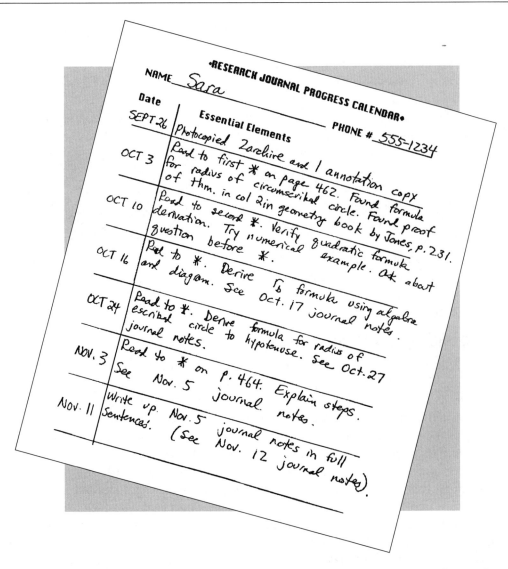

•RESEARCH JOURNAL PROGRESS CALENDAR•

NAME Sara

PHONE # 555-1234

Date	Essential Elements
SEPT 26	Photocopied Zarchive and 1 annotation copy
OCT 3	Read to first ✱ on page 462. Found formula for radius of circumscribed circle. Found proof of thm. in col 2 in geometry book by Jones, p. 231.
OCT 10	Read to second ✱. Verify quadratic formula derivation. Try numerical example. Ask about question before ✱.
OCT 16	Read to ✱. Derive r_b formula using algebra and diagram. See Oct. 17 journal notes.
OCT 24	Read to ✱. Derive formula for radius of escribed circle to hypotenuse. See Oct. 27 journal notes.
NOV. 3	Read to ✱ on P. 464. Explain steps. See Nov. 5 journal notes.
NOV. 11	Write up Nov. 5 journal notes in full sentences. (See Nov. 12 journal notes).

tation depends on the questions and material that need to be discussed. Before each consultation, read parts of your article and be prepared to discuss what you've read. The amount of time you have available, the natural breaks in the article's flow, and your teacher's advice should all be considered when deciding how much to read between consecutive consultations. A week's reading could consist of one page, one column, or one paragraph.

Mastery of the reading selection and extension of it will take up much of your time, because these articles are written for people who are college math majors or math teachers. Part of your job will be processing the information and explaining it more compreively in your paper. Most of the articles will be three to six pages long. Theoretically, you could read the words on six pages in a few minutes. Before you begin your slow, careful, critical reading of an article, take time to skim it. Look at the section titles and the major theorems, diagrams, and tables. Read the conclusion. Make some mental notes about what looks familiar and what doesn't. After familiarizing yourself with the article, begin to read it more critically. Your mastery of the material will come as you read slowly, so don't proceed until you have mastered what you have already read. In some cases, you will need to read single words, sentences, and paragraphs several times before you understand them. Take time to think at the end of each paragraph or sentence. Digest the informa-

tion, and then think about possible questions, examples, conjectures, and additional material that relate to the reading passage.

Reading and Note Taking

As you read, annotate your copy of the article by highlighting key words and phrases and by indicating where you have a question. You should test any claims that the article makes with examples of your own. Learn to read between the lines in mathematics. Because journal articles are written for people with deeper mathematical backgrounds than your own, steps are left out of proofs, and sometimes entire proofs are left out. Examples and diagrams are omitted and tables are condensed to save space in the article. Your finished paper will present the material in the article, along with original material, in a document that can be read and understood by a high school student with greater ease and more depth than could the original article. Therefore, you have to supply the information that is implied "between the lines." Always read with paper, pen, and highlighter in hand. You'll need to take notes that don't fit in the margins of your annotation copy. These notes will make your research journal grow quickly. Don't throw away notes—you might throw away something that could be of value later on. Don't forget to keep conjectures that you eventually prove to be incorrect.

We have spoken in general terms about some reading and note-taking skills. Now we take a look at a specific article, "On the Radii of Inscribed and Escribed Circles of Right Triangles," by David W. Hansen. This three-page article appeared in the September 1979 issue of *Mathematics Teacher*. We show the article here, complete with Sara's annotations.

ON THE RADII OF INSCRIBED AND ESCRIBED CIRCLES OF RIGHT TRIANGLES

Handwritten note (left margin): Possible extension to equilateral? Special case of isosceles right triangles?

Mixing the Pythagorean theorem, the area of a triangle, and some first-year algebra yields some unexpected results.

By DAVID W. HANSEN
Monterey Peninsula College
Monterey, CA 93940

An *inscribed* circle of a triangle is a circle tangent to the three sides of the triangle with its center inside the triangle. An *escribed* circle of a triangle is a circle tangent to one side and to the extensions of the other two sides with its center outside the triangle. Every triangle thus has one inscribed and three escribed circles (see fig. 1). If we restrict our interest to right triangles only, an interesting relationship between the radii of the inscribed and escribed circles can be found.

Handwritten note (left margin): Also, one circumscribed circle. Find its radius.

Consider the radius of the inscribed circle of right triangle ABC as shown in figure 2. Let the lengths of the sides of the triangle be a, b, and c, with G the center of the inscribed circle, r its radius, and D, E, and F the points of tangency of the sides of the triangle with the inscribed circle G. Then, since a circle's radius is per-

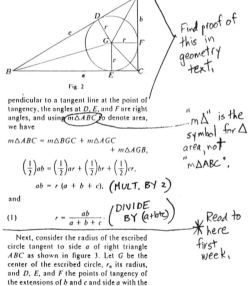

Fig. 2

Handwritten note: Find proof of this in geometry text.

pendicular to a tangent line at the point of tangency, the angles at D, E, and F are right angles, and using $m\triangle ABC$ to denote area, we have

$$m\triangle ABC = m\triangle BGC + m\triangle AGC + m\triangle AGB,$$

$$\left(\tfrac{1}{2}\right)ab = \left(\tfrac{1}{2}\right)ar + \left(\tfrac{1}{2}\right)br + \left(\tfrac{1}{2}\right)cr,$$

$$ab = r(a + b + c). \quad \text{(MULT. BY 2)}$$

Handwritten note: "$m\triangle$" is the symbol for $\triangle$ area, not "$m\triangle ABC$".

and

$$(1) \qquad r = \frac{ab}{a+b+c} \qquad \left(\begin{array}{l}\text{DIVIDE}\\\text{BY } (a+b+c)\end{array}\right)$$

Handwritten note: ✱ Read to here first week.

Next, consider the radius of the escribed circle tangent to side a of right triangle ABC as shown in figure 3. Let G be the center of the escribed circle, r_a its radius, and D, E, and F the points of tangency of the extensions of b and c and side a with the escribed circle G. As before, the angles at the points of tangency are right angles, and

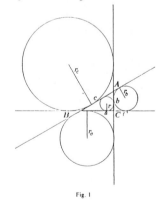

Fig. 1

STATE IN FULL, Find proof.

$\triangle ADG \cong \triangle AEG$, with $\triangle BFG \cong \triangle BEG$ by the hypotenuse-leg <u>congruence theorem</u> for right triangles. Thus,

$$m\triangle ABC + m\triangle FCDG + m\triangle BEG + m\triangle BFG = m\triangle ADG + m\triangle AEG.$$

develop in gradual stages

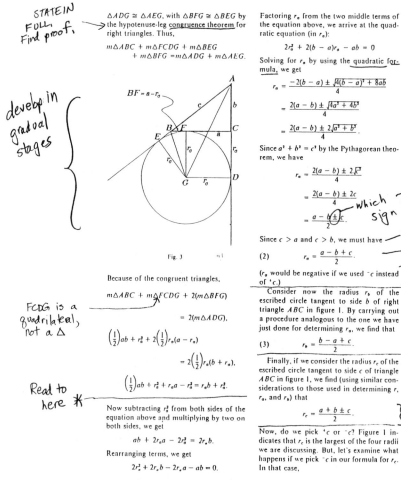

Fig. 3

Because of the congruent triangles,

FCDG is a quadrilateral, not a △

$$m\triangle ABC + m\triangle FCDG + 2(m\triangle BFG)$$
$$= 2(m\triangle ADG),$$
$$\left(\tfrac{1}{2}\right)ab + r_a^2 + 2\left(\tfrac{1}{2}\right)r_a(a - r_a)$$
$$= 2\left(\tfrac{1}{2}\right)r_a(b + r_a),$$
$$\left(\tfrac{1}{2}\right)ab + r_a^2 + r_a a - r_a^2 = r_a b + r_a^2.$$

Read to here ✳

Now subtracting r_a^2 from both sides of the equation above and multiplying by two on both sides, we get

$$ab + 2r_a a - 2r_a^2 = 2r_a b.$$

Rearranging terms, we get

$$2r_a^2 + 2r_a b - 2r_a a - ab = 0.$$

Factoring r_a from the two middle terms of the equation above, we arrive at the quadratic equation (in r_a):

$$2r_a^2 + 2(b - a)r_a - ab = 0.$$

Solving for r_a by using the <u>quadratic formula</u>, we get

$$r_a = \frac{-2(b - a) \pm \sqrt{4(b - a)^2 + 8ab}}{4}$$
$$= \frac{2(a - b) \pm \sqrt{4a^2 + 4b^2}}{4}$$
$$= \frac{2(a - b) \pm 2\sqrt{a^2 + b^2}}{4}.$$

→ $x = \dfrac{-b \pm \sqrt{b^2 - 4ac}}{2a}$ warn readers about not confusing a's and b's and c's.

Since $a^2 + b^2 = c^2$ by the Pythagorean theorem, we have

$$r_a = \frac{2(a - b) \pm 2\sqrt{c^2}}{4}$$
$$= \frac{2(a - b) \pm 2c}{4}$$
$$= \frac{a - b \pm c}{2}.$$

which sign? WHY?

CAN THIS BE CONSTRUCTED USING EUCLIDEAN TECHNIQUES?

Since $c > a$ and $c > b$, we must have

(2) $$r_a = \frac{a - b + c}{2}.$$

WHY?

(r_a would be negative if we used $-c$ instead of $+c$.)

✳ Read to here

Consider now the radius r_b of the escribed circle tangent to side b of right triangle ABC in figure 1. By carrying out a procedure analogous to the one we have just done for determining r_a, we find that

(3) $$r_b = \frac{b - a + c}{2}.$$

✳ Read to here and derive

Finally, if we consider the radius r_c of the escribed circle tangent to side c of triangle ABC in figure 1, we find (using similar considerations to those used in determining r, r_a, and r_b) that

$$r_c = \frac{a + b \pm c}{2}.$$

derive

Read to here

Now, do we pick $+c$ or $-c$? Figure 1 indicates that r_c is the largest of the four radii we are discussing. But, let's examine what happens if we pick $-c$ in our formula for r_c. In that case,

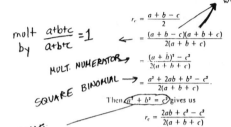

the difference of 2 squares, $x^2 - y^2$, where $x = a+b$ and $y = c$.

mult $\dfrac{a+b+c}{a+b+c} = 1$ by

MULT. NUMERATOR

SQUARE BINOMIAL

$$r_c = \frac{a+b-c}{2}$$
$$= \frac{(a+b-c)(a+b+c)}{2(a+b+c)}$$
$$= \frac{(a+b)^2 - c^2}{2(a+b+c)}$$
$$= \frac{a^2 + 2ab + b^2 - c^2}{2(a+b+c)}.$$

Then $a^2 + b^2 = c^2$ gives us

SINCE THESE ARE RIGHT TRIANGLES

$$r_c = \frac{2ab + c^2 - c^2}{2(a+b+c)}$$
$$= \frac{2ab}{2(a+b+c)}$$

or $\quad r_c = \dfrac{ab}{a+b+c}.$

But equation (1) tells us that the radius of the inscribed circle r equals

$$\frac{ab}{a+b+c}$$

Now, this is impossible. Thus, ^+c must be selected in order to give the radius of the escribed rather than the inscribed circle. Thus, we have

read ✳

(4) $\quad r_c = \dfrac{a+b+c}{2}$

Summarizing our results, we have found that

$$r = \frac{ab}{a+b+c} = \frac{a+b-c}{2}$$
$$r_a = \frac{a-b+c}{2}$$
$$r_b = \frac{b-a+c}{2}$$
$$r_c = \frac{a+b+c}{2}.$$

Looking at the results above, we see a beautiful pattern of symmetry. Each radius equals one-half the sum or difference of the three sides of the triangle. Furthermore, r, r_a, and r_b each involve one subtraction, whereas r_c involves only addition, implying that r_c is larger than any of the other radii. But we can say more. By adding r, r_a, and r_b, we get

$$r + r_a + r_b = r_c.$$

derive this algebraically

Thus, we find that the sum of the radius of the inscribed circle and the radii of the two escribed circles tangent to the legs of a right triangle is equal to the radius of the escribed circle on the hypotenuse of the right triangle.

$$r_c = r + r_a + r_b.$$

But this is not all. If we multiply the smallest radius r by the largest radius r_c, we find that

derive this algebraically

$$rr_c = \frac{ab}{2}.$$

which is the area of $\triangle ABC$. Also,

$$r_a r_b = \frac{ab}{2},$$

derive this algebraically

again the area of $\triangle ABC$.

Thus, we find that the area of a right triangle is equal to either (1) the product of its inscribed radius and the radius of the escribed circle on its hypotenuse or (2) the product of the two radii of the escribed circles on its two legs.

$$m\triangle ABC = rr_c = r_a r_b.$$

The calculations below illustrate our findings for a right triangle with sides 3, 4, and 5 (see fig. 1). Let $a = 4$, $b = 3$, and $c = 5$, then

$$r = \frac{4+3-5}{2} = 1$$
$$r_a = \frac{4-3+5}{2} = 3$$
$$r_b = \frac{3-4+5}{2} = 2$$
$$r_c = \frac{3+4+5}{2} = 6.$$
$$r_c = r + r_a + r_b = 1 + 3 + 2 = 6$$
$$rr_c = (1)(6) = (3)(2) = r_a r_b = 6$$

and the area of $\triangle ABC$ is $(1/2)(4)(3) = 6$.

MAKE OTHER EXAMPLES, some with irrational sides.

BIBLIOGRAPHY

Ogilvy, C. Stanley, and John T. Anderson. *Excursions in Number Theory.* New York: Oxford University Press, 1966.

Long, Calvin T. *Elementary Introduction to Number Theory.* Boston: D. C. Heath & Company, 1965.

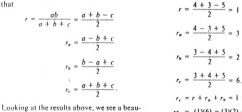

Two old books make up bibliography tree.

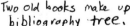

Read next article "Pythagoras Meets Fibonacci" from April 1981 *Math Teacher*. Combine results if possible.

Let's examine some of Sara's annotations. Notice that the word *right* is circled in the title. The article is deliberately limited to claims about right triangles, and skimming over this single important word could mislead you. Throughout the article, notice the words and phrases Sara underlined. Can you tell why she underlined them? Keep in mind that even a single word or a short phrase can be significant. Sara wanted to call attention to words she considered important.

Before we continue examining Sara's annotations, let's look at how leaving out or adding a word can drastically change the meaning of a passage. How crucial can one word be? Read the following sentence and draw diagrams to test the claim. Determine whether the statement is true or false.

If two triangles are similar and two unequal sides of the first triangle are congruent to two sides of the other triangle, the triangles are congruent.

The statement is false. The triangles are not necessarily congruent, because the pairs of congruent sides are not necessarily *corresponding* sides.

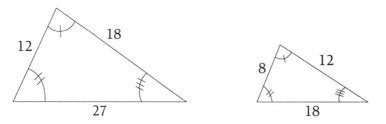

Similar triangles with two pairs of congruent sides.

The triangles above are not congruent. If pairs of corresponding sides were equal, the triangles *would* be congruent. Readers who answer true usually inferred the word *corresponding*. The meaning of the statement is drastically altered by the inadvertent mental insertion of this word. As you can see, leaving out or adding one single word can lead to confusion. If Sara hadn't noted that the article dealt only with *right* triangles, she might have wasted a lot of time trying to apply the findings in the article to nonright triangles. Now you can see why Sara underlined so carefully.

Sara's knowledge of geometry led her to want to inquire about the circumscribed circle, as annotated in the first paragraph. In column 2 of page 462 of the article, there is a theorem for which she will find a proof. In the middle of this column, she notes an unfamiliar and potentially misleading notation for area. Notice that Sara's first reading assignment ended at the asterisk in column 2. She will address some of the points she raised before she continues her reading.

On page 463, Sara made some more notes. You can see that this page alone was divided into four different reading assignments. These were apportioned by Sara in a consultation with her teacher. Several claims are made that Sara must test and derive. At the top of column 1 is a theorem that Sara must state in full in her formal paper, so she will find the theorem and its proof before her next reading session. Notation for area in the middle of the column seems to label quadrilateral *FCDG* as a triangle. This never confused Sara, because she knew the intent. Sara plans to develop the diagram in column 1 in stages, using several diagrams with explanations in between each. She is also going to try to construct the radii of the circles using compass and straightedge. Toward the end of the page, claims are made that Sara must prove. Here she must read between the lines and supply original examples and derivations not given by the author.

Page 464 shows more annotations that again require reading between the lines. All of the information Sara compiles becomes part of her research journal. As she accumulates information, she makes sure it is comprehensive and legible, so it can easily be incorporated into her formal paper at a later date. A sample page from Sara's journal follows on the next page. This page is her original derivation of a claim made on page 464, column 2. Take time to look at the sample page before continuing. Notice that this journal entry is clear and complete. Sara will have no trouble relearning this material when it is time for her to write this section of her paper. (See opposite page.)

Sara could not find the books listed on her bibliography tree, so she looked for other articles. She found one on the Fibonacci sequence and the Pythagorean theorem. After looking at the article, she decided she could try to use some of the results to make new generalizations about special right triangles and escribed circles. The article showed how to generate Pythagorean triples using the Fibonacci sequence. Sara thought that if she examined right triangles with integer-length sides generated from the Fibonacci sequence, she might be able to express the radii of the circumscribed, inscribed, and escribed circles in terms of the original Fibonacci numbers used. Sara read and annotated this article and then attempted to combine the findings from both articles. Can you see how the original material Sara added to the first article and the combination of results with a second article made her paper a research paper, not just a math report? She began the formal write-up after two to four months of research. Her research continued during the writing process.

Conquering Difficult Concepts

Some parts of your articles will be more difficult to understand than others. If you don't agree with a finding, you might suspect that there is an error in the article. Notice that Sara's article used unconventional notation for area. Generally speaking, journal articles are edited very carefully, and it is rare to find a mistake in an article. However,

Sara L.   Dec 2

Derivation of claim from page 464, column 2 :
The area of the right triangle is equal to
the product of the radii of the two smaller
escribed circles.

$$r_a = \frac{a-b+c}{2} \qquad\qquad r_b = \frac{b-a+c}{2}$$

$$r_a \cdot r_b = \left(\frac{a-b+c}{2}\right)\left(\frac{b-a+c}{2}\right)$$

$$= \frac{ab - a^2 + ac - b^2 + ba - bc + cb - ca + c^2}{4}$$

$$= \frac{2ab - a^2 - b^2 + c^2}{4}$$

$$= \frac{2ab - \overbrace{(a^2+b^2)}^{=c^2} + c^2}{4}$$

$$= \frac{2ab}{4} = \frac{ab}{2} = \text{Area of Right } \triangle$$

Sample page from Sara's research journal.

sharp students reading carefully and testing all claims have uncovered some errors.

> Robin was reading an article on the Malfatti Problem in the May 1992 issue of *Mathematics and Informatics Quarterly*. An inequality expression erroneously had the inequality sign reversed. Robin wrote to the editor and received a reply that, indeed, there was an error. Robin had spent hours trying to understand the incorrect inequality!

> Tim was reading an article on the arbelos in the April 1973 issue of *Mathematics Teacher*. An equation erroneously had a factor of 1/2 in it. Tim found the error based on a diagram in the article and the definition of an arbelos, and he didn't base any further research on the error.

Many times, readers who *think* they have found an error don't fully comprehend what they are reading. Upon rereading, trying examples, and so on, readers usually find that there are no errors. Occasionally, you may find an error, but don't assume that material you disagree with is in error. Read again carefully, and try some examples and diagrams to see if you can eliminate the problem. A cooperative effort might clear up the discrepancy. If another student is working from the same article, the two of you should get together and discuss the problem. If there are no other students working on your article, decide whether you can isolate the problem and explain its context to another student or to your teacher. Another person doesn't necessarily need to be familiar with the entire article to be able to help you out.

It is possible to move ahead with your reading even if you have questions about previous material. You might decide to accept the result you have questions about and read ahead, keeping in mind that the readings are based on the truth of the result you don't fully understand. Persistence, your problem-solving strategies, and your newfound reading and journal-keeping skills will go a long way in helping you decipher challenging material. Discussing the material at a consultation with your teacher will also help you master the tougher parts of your article.

The Journal Article Reading Assignment

In Chapter 4, you completed a Math Annotation Project to gain experience in writing mathematics. This experience will help you take notes for your journal and write your formal research paper. Now that you have completed this chapter, you need experience in reading a mathematics article. Before you begin reading the research article you have selected for your paper, you should read a different article with a group of students, preferably your whole class. Annotate the article, take notes, list questions, and so on as described in this chapter. You should discuss the article as part of a group, taking notes on the observations of your classmates. This Journal Article Reading Assignment is a primer for the readings that will be part of your research. Take advantage of the fact that you can discuss the Journal Article Reading Assignment with other students—the feedback is very valuable.

At this point, you know that your research journal will consist of questions, examples, conjectures, theorems, patterns, derivations, proofs, and counterexamples. Over a period of months, you will add material to your research journal. All of this information will be compiled and presented in a logical fashion in a formal research paper that combines your research and your writing skills. Chapter 8 will help you assemble the components of your final paper.

Chapter Eight
Components of Your Research Paper

When you write mathematics, your aim is to convey information as clearly and effectively as possible for a limited, specialized audience that we call the target audience. The target audience must have the requisite background to read the research. Precision, completeness, and clarity are paramount in mathematical writing—it can never be too clear. A finished mathematical piece will not read like a novel or a magazine article, and rightfully so. The purpose and the style are different. The poetic license you use as you create high-quality English essays is inappropriate in a math research paper. This is not to say that mathematical research papers cannot be enticing, riveting, or even suspenseful. You can arouse the curiosity of the reader by posing questions at the beginning of your paper that you will answer in your paper. The mission of the written paper is the transmission of newly-found results and the ability to provoke the search for new knowledge. To the mathematician, the beauty of the findings is pure poetry. A research writer has succeeded if a member of the target audience becomes "hooked" by the intrigue of your opening questions and then can read, understand, and intelligently discuss the research. If your presentation is clear, future researchers can use it as a basis for further investigation in the field.

The writing and reading you have done all your life have contributed to your present writing ability. You will need to tap all the skills you have acquired as you put your paper together. In this chapter, we discuss the specific parts of your research paper. You should get a copy of a writing handbook from your school's library or English department. Use this handbook as a reference throughout the entire writing process. Keep in mind that formal writing follows sets of rules just as algebra does. Your research journal and your own memory will provide you with the mathematics material you need to build your paper.

The Structure of Your Research Paper

Before you write, you need a plan—you need to write an outline of your research paper. Before writing this outline, however, you should become familiar with the basic parts of your research paper, because your outline will be built around them. They are listed here in the order in which they will appear in your finished paper. You might not write the items in this order; we discuss why later in this chapter.

I.   Cover Page
II. Abstract
III. Problem Statement

IV. Body of the Paper
V.   Recommendations for Further Research
VI. References
VII. Appendix

We will discuss the function of each component of the paper in detail. Then we will examine a sample outline for a research paper.

Cover Page

The cover page features the title of your paper. You might think of a title as you are doing your research, and you might revise this title several times as your work progresses. Why? The title should be descriptive; it may be longer than titles you've written for reports in other contexts. For example, Sara read the article, featured in Chapter 7, on escribed circles to right triangles. She also read a second article that she notes in her annotations at the end of the first article. This second article describes how to use Fibonacci numbers to create Pythagorean triples. After completing the research of each article, Sara tried to combine results from the two articles and came up with a new, interesting result. She made this result the focus of her research, and it is reflected in her title. The title, as written in inverted pyramid form on her cover page, is shown here:

THE RELATIONSHIP BETWEEN THE SIDES OF A RIGHT TRIANGLE
GENERATED BY FIBONACCI NUMBERS AND THE
RADII OF ITS ESCRIBED CIRCLES

The following titles would not have been as informative to the reader, because they are not as descriptive:

RIGHT TRIANGLES AND THEIR CIRCLES
or
CIRCLES, RIGHT TRIANGLES AND FIBONACCI NUMBERS

Which title below do you think is more informative for the potential reader?

DERIVING THE EUCLIDEAN METHOD OF GENERATING PRIMITIVE
PYTHAGOREAN TRIPLES
or
THE PYTHAGOREAN THEOREM

The title appears in inverted pyramid form on the cover page. The cover page is also the place for your name, your affiliation, and the date of your paper. A sample cover page is shown in Appendix B_1. The title itself should reflect the problem statement, because that is the major focus of the research. Although your cover page will, in most cases, be written after your paper is completed, keep in mind that it is linked to the problem statement.

Problem Statement

The first section of a research paper is titled "Problem Statement." It includes an introduction that succinctly orients the reader to your topic. Make sure your introduction is appropriate for the target audience. The reader needs some previous knowledge to follow your paper but should not become immersed in the complex specifics of your research until the stage has been set. Appendix B_3 has a sample of the Problem Statement section of a student's research paper. Note that the problem is stated early in the research paper. Although you must orient the reader, you should not give an excessive amount of background material before stating the problems. The questions posed in the Problem Statement section are the problems you will address in your paper. They could be problems posed in the article you read, original problems, or a combination of both. Read the example in Appendix B_3 and determine how well you

understand what the papers will be addressing. You can read other examples of problem statements in the sample outline toward the end of this chapter.

Body of the Paper

In Chapters 3 and 4, you learned about the formal writing of mathematics. Tips, examples, and suggestions were given. Hopefully, you had a chance to practice some of these writing suggestions as part of a Math Annotation Project. (Your research journal entries may have been written more informally, since they were notes to be read by you only.) Use Chapters 3 and 4 as references as you write the body of your paper, along with the suggestions given here.

❑ Use headings to divide the paper into different sections at logical junctures. Headings were first discussed in Chapter 3. We review them here because they reflect parts of the research paper and are not as open-ended as the headings you created for the Annotation Project. Your outline can serve as a skeleton for constructing your paper. Devise a hierarchy as to how you will indicate new sections and divisions within sections. For example, you can use the techniques of justification, lower- and uppercase letters, and underlining to create a multilevel hierarchy of headings in your paper. Here is one possibility:

SECTION HEADINGS ARE CENTERED, ALL UPPERCASE

FIRST SUBSECTION HEADINGS ARE LEFT JUSTIFIED, UPPERCASE

Next Subdivisions of Sections Are Centered, Mixed Upper- and Lowercase

<u>The Next Heading Is Left Justified, Mixed Upper- and Lowercase, and Underlined</u>

<u>Another Heading Can Be Indented, Mixed Upper- and Lowercase, and Underlined</u>

For example, let's look at a sample subdivision of Sara's paper. Keep in mind that the cues for each new section came from the structure of her outline. (Her outline is examined in the next section. After you have read the section about creating an outline, compare her outline with her subdivisions.)

PROBLEM STATEMENT

RELATED AND ORIGINAL RESEARCH

TRIANGLES AND THEIR CIRCLES

The Radii of the Circumscribed and Inscribed Circles

The Radii of the Three Escribed Circles

<u>The Escribed Circle to Side *a*</u>

<u>The Escribed Circle to Side *b*</u>

<u>The Escribed Circle to the Hypotenuse</u>

RIGHT TRIANGLES AND FIBONACCI NUMBERS

Generating Sides from Consecutive Fibonacci Numbers

Formulas for Radii in Terms of Fibonacci Numbers

<u>The Radius of the Circumscribed Circle</u>

<u>The Radius of the Inscribed Circle</u>

<u>The Radii of the Three Escribed Circles</u>

RECOMMENDATIONS FOR FURTHER RESEARCH

REFERENCES

APPENDIX

Headings should never be placed at the bottom of a page; if the word processor breaks a page at a heading, move the heading to the top of the next page. If you have a consistent system for dividing up your paper, readers will clearly understand how the different sections and subsections of your research are related. There is a close relationship between the outline and the section divisions of your paper.

❏ Within each section, divide your paper into paragraphs pragmatically with respect to mathematical content development. Remember to center and highlight mathematical expressions that need to be "digested." Recall from Chapter 3 that the page layout of your paper will be visually different from that of a novel. Present your material with logical breaks indicated by new paragraphs. Make sure your paragraphs and sections are logically connected to those that follow. Use sentences (segues) at the end and beginning of sections that explain why "headings" occur at their specific juncture in the paper. Whenever possible, end sections, subsections, and paragraphs with a motivating idea that is addressed by the next section or paragraph.

For example, suppose you were writing a paper on the quadratic formula. You would discuss roots, factors, trinomials, and binomials and possibly even show a geometric interpretation of some of your major points. You would start with trinomials that were factorable. If section three of your paper develops a derivation of the quadratic formula, you can end section two with your paper's first mention of a quadratic equation with a trinomial that can't be factored, such as

$$x^2 + 8x + 6 = 0$$

After having "indulged" the reader in solving quadratic equations by factoring, you have now motivated the need for another method, *because factoring doesn't work!* After reading and understanding your research on the factorable equations, the reader *wants* to know how to solve the problem you've posed. Now your treatment of the quadratic formula is not arbitrary but driven by need, creating a solid flow between these two sections of your paper. When you proofread, determine whether your paragraphs and sections flow sensibly.

❏ Credit authors if you use their findings, even if you use no direct quotes. These credits are called **citations**. Include the author's name and year of publication in any citation. If interested, your reader can find complete information about a source in the References section of your paper. You should look up citations in your writing handbook. Here are some examples:

Trobiano (1995) found that any whole number can be expressed as . . .
or
In 1995, Trobiano found that any whole number can be expressed as . . .
or
Any whole number can be expressed as . . . (Trobiano, 1995).

❏ Follow the rules you learned in English class for direct quotations, or use a writing handbook. Make sure you include at least the author's last name and the year the material was published. If interested, your reader can find complete information about a source in the References section of your paper. Short quotes usually require quotation marks; longer quotes (forty or more words) are sometimes indented in their own block of text, without quotation marks.

Recommendations for Further Research

Recall from Chapter 1, or from your own research experience, that each new venture in mathematics raises many questions. In your research you found some answers, and you may have wondered about the answers to questions that arose in relation to your research. These questions might fall outside the narrow focus of your research, but they can be posed for other researchers to pursue. They are logical next steps based on the conclusions made in your paper and could actually become the problem statements for another research paper. Appendix B_9 shows an example of Recommendations for Further Research. This concluding component of your paper is *not* a summary of your findings. Notice that it is a short section.

Abstract

When your paper and cover sheet are completed, you can write an **abstract**, a short summary of what the reader can expect to find in your paper. It appears right after the cover page. An abstract is not an introduction—it is a summary. It should be approximately 250 words or less, on a separate page headed "ABSTRACT." The abstract should contain one or two paragraphs and be single-spaced, centered between the top and bottom margins. It shouldn't go into great detail, but make sure it conveys the essence of your research. An abstract allows future researchers to determine quickly whether your research might be of interest to them. Below are two examples of students' abstracts. The first is Tim's abstract from his paper on Bobillier's theorem.

Bobillier's theorem states that "if a triangle of fixed size moves in the plane in such a way that two of its sides are tangent to two fixed circles, then the third side will be tangent to another fixed circle." This research develops a proof of the theorem. Relationships between circles and tangents, and circles and triangles, are explored to give an overview of the concepts necessary to understand the proof of the theorem. A lemma (a theorem whose primary function is in the proof of another theorem) for Bobillier's theorem is proven, making use of these concepts, and the theorem itself is proven. Suggestions for further research concerning the theorem are included at the end. These include possibly using other polygons and nonconvex polygons in place of a triangle.

Daniele's research on probability and the quadratic formula is summarized in her abstract.

This research explores a purely mathematical question and analyzes its solution. This research investigates the probability that a quadratic equation $y = ax^2 + bx + c$, where b and c are randomly selected real numbers and $a = 1$, has real roots. The problem is approached through a number of different methods and techniques, some more accurate and efficient than others. One method is the Monte Carlo method, which uses area relationships to find probabilities. The probability of the roots being real is verified using integral calculus. In finding the solution, the research relies on conic sections, calculus, functions, geometry and probability.

Another sample abstract can be found in Appendix B_2. Because the abstract is a summary, writing it is the last step in writing the research paper. The cover page and the abstract establish that the research to follow is serious, high-quality material.

References

The end of your paper will need a reference list, or bibliography. In many cases, this will include only the articles you read. You must include any article that you cite in your paper. There are several different styles that your references can follow. Use the style you were taught in English class, or adopt another form from a writing handbook. Your librarian may also be able to help you. You can follow the form used in the reference sections of the books and articles you've read. Start by taking a look at the bibliography of this book.

Appendix

Under certain circumstances, your paper will need an Appendix. The Appendix appears at the end of the paper and houses materials that should not appear in the body of the paper, mainly due to their length. What information might comprise your Appendix? You may have used a large set of raw data in your paper. If you ran information on a computer, you might want to include a copy of the program and the output. If your research involved the use of forms, questionnaires, or surveys, include these in your Appendix. If you corresponded with the author of an article, include all letters. If you interviewed someone for your research, you can include a transcript of the interview. Pertinent newspaper and magazine clippings can also appear in the Appendix. The Appendix is the correct place for items that are too cumbersome for the body of the paper, such as lengthy lists. You can refer to the Appendix in the body of your paper. Not all papers will have an Appendix.

Creating an Outline

The outline of your research paper should present the material in a logical order. Often the order is based on the order of presentation in your readings. Your readings may include claims and proofs that skip steps or combine steps, because the intended audience of the article is someone with a mathematics background. You can add some original statements, reasons, and examples to improve the article's "abridged version." (We previously called this "reading between the lines.") You will have to decide where your extensions and "between the lines" original material fit. Be sure to include all of this material in your outline. You will need to revise your outline if, during the course of your work, you decide to change the material you present or the order of its presentation.

Let's look at a sample outline for a research paper. This outline is based on the article in Chapter 7 that was read and annotated by Sara. Notice from her outline that Sara also read a second article, on Fibonacci numbers and right triangles. If you would like to compare her outline to the actual article in Chapter 7, keep in mind that items IC4, IIC, IID, IIIB, and IIIC refer to the second article she read. These items would not be included in a paper based solely on the escribed circles article in Chapter 7. The remainder of the outline *would* be appropriate for such a paper.

I. Introduction & Problem Statements
   A. Note limitation to right triangles
   B. Define circumscribed, inscribed and escribed circles
   C. Problem statements
      1. What is the relationship between the sides of a given right triangle and the radius of its circumscribed circle?
      2. Same as #1, for inscribed circle?
      3. Same as #1, for each of its 3 escribed circles?
      4. If the sides of a right triangle are generated by Fibonacci numbers as described in the article

by Boulger (1989), what is the relationship between the radii of the escribed circles and the original Fibonacci numbers?

II Body of paper
  A. Derive formulas for circumscribed & inscribed radii - journal Oct. 7
  B. Escribed circles' radii
     1. Radius of circle to side 'a' - see journal Oct. 12
        a. Show Euclidean construction of radius - journal Oct. 15
     2. Radius of circle to side 'b' - analogous derivation - journal Oct. 21
        a. Euclidean construction
     3. Original derivation of radius of circle to hypotenuse
        a. Euclidean construction - see journal Oct. 30
  C. Fibonacci article
     1. Give formulas for sides of $\triangle$; cite article
        a. Show formulas satisfy $a^2 + b^2 = c^2$; journal Nov. 30
        b. Numerical examples - see journal Nov. 30
  D. Combining results from the two articles
     1. Radius to side 'a' in Fibonacci terms - see journal Dec. 7
     2. Same for side 'b'. See journal Dec. 20
     3. Hypotenuse - see journal Jan 7
     4. Examples using spreadsheet

III Recommendations for further research
  A. Find radii for escribed circles to equilateral $\triangle$.
  B. Relationship between right $\triangle$'s area & Fibonacci #'s
  C. Relationship between right $\triangle$'s perimeter & Fib. #'s

IV Appendix, References, Cover Page, Abstract

Sample research paper outline

Putting the Parts Together

When you have completed your outline, consult your journal and start writing your introduction. As you complete pages of your writing, you will read them over and review them. You can avoid submitting drafts with many errors by using the suggestions in this chapter and Chapter 3 as a checklist when you proofread and edit your work. Appendix B has additional student sample pages you can use for guidance. Proofread and edit every draft before submission. Address all of your teacher's comments from previous drafts. Your teacher may have some examples of former or in-progress student papers you can look over.

As your work progresses throughout the year, you will develop valuable research skills, including reading, writing, conjecturing, problem solving, and proving theorems. If, during the course of your high school career, you write more math research papers, you will grow in your experience of mathematics as an inductive, discovery-driven science. You might decide to enter some of the mathematics contests available to high school students. If you are interested in entering your research paper in a contest, read the next section and discuss your options with your teacher, math department chairperson, or principal.

Entering Math Contests

Your experience with reading, writing, and research skills has made you a better researcher. These skills will also help you become a better problem solver. You are now better equipped to read problems, extend problems, and write well-explained solutions to problems. If you enjoy problem solving and/or math research, there are several contests open to high school students just like you!

Contests can be schoolwide, countywide, statewide, national, or even international. They may involve problem solving or writing a research paper. Some require teamwork, and others require individual entries. Some require an extensive written solution to a problem, which is sometimes open-ended. The solution may even require that you build

something. Some contests are in the form of scholarship exams. Contests will tap every facet of your problem-solving and math research ability. They provide a unique forum for you to blend communication, reasoning, connections, problem solving, and technology with mathematics. Prizes for contest winners range from plaques to trophies, trips, books, cash, and even college scholarships. Not everybody wins a contest but in many ways all entrants are winners. When you complete a project that really tested your mettle, made you work to your potential, and instilled knowledge in you, you truly are a winner!

Each year there are many contests, some new and some perennial favorites. At certain times, some contests may be discontinued. To keep abreast of the currently available contests for high school students, you can write to the organizations listed in Appendix A. Contact your local and state mathematics teachers associations for a list of contests in your area. Many contests that, by name, appear to be *science* contests include mathematics as a science. A list of some major contests is given in Appendix C_4. In some cases, you can apply for information directly. In other cases, your teacher should send for the information. Talk to your teacher about securing information about each contest; then you can decide which are for you.

Read the rules carefully for any contest you plan to enter. Contests involving research papers may have spacing requirements and a page minimum and/or maximum. There could be other regulations that require you to go back to the word processor and adjust your paper to conform to the required form. Don't disqualify a fine effort because you didn't follow directions. Read and scrutinize!

If the prospect of a contest deadline seems a bit overwhelming now, prepare for the future. Send for the application packets and become familiar with the expectations of each contest. You may decide to enter at another time. You might work on your research paper this year and decide to continue it next year. The summer is a great time to make headway on research papers. Many universities offer summer programs for high school students, and many of these programs provide support services for students involved in research. Contact your local colleges to find out if they offer such a program. For information on national and international programs, look in the journals and newsletters of mathematics and education organizations or write to them requesting information on summer programs. Also speak to your school's guidance department.

Some contests require the entrant to make an oral presentation in addition to submitting written work. Chapter 9 will help you put together an oral report that reflects the high quality of a well-done research paper.

Chapter Nine
Oral Presentations

At certain times in your life, you will be judged not solely on your credentials or on your written work but on your performance at a "live audition." Certainly, a job interview is a classic case of this. You send in a resume, school transcripts, and references for a job, all in the hope of gaining an interview. If you are granted an interview, then it is must be that you are qualified, on paper, for the job. The interview allows a potential employer to see how you carry yourself, that is, how you answer questions and handle discussions. A similar scenario may take place in college interviews. Many people are nervous in an interview situation.

Public speaking also makes some people nervous. Have you ever given a speech to a large group? Most people don't do this often and therefore don't get much practice at it. Presenting scholarly material requires knowledge as well as an effective spoken delivery. You will have to answer questions. Thinking on your feet, without a prepared script, makes giving a good live academic presentation both a challenge and a very rewarding experience. It is difficult to simulate the exact conditions of such a presentation, so most practice sessions are done without an audience. If you plan to make an oral presentation on your math research, what steps can you take to make sure you look professional, even if you've had little practice speaking in front of a group? This chapter will help you methodically prepare for your oral presentation so that you can look forward to it with confidence.

P⁴—*P*lanning, *P*reparing, and *P*racticing the *P*resentation

Your math research paper represents months of work. You may already know much of it off the top of your head because you have internalized it by reading carefully, discussing the material at consultations, working through original material, and frequently revising your written presentation. A command of the material in your paper is the first step in building your oral presentation. With this as a foundation, your next goal is getting your point across to your audience. Start by planning your oral presentation.

Planning Your Oral Presentation

Two key factors will affect how your research is presented. First, you must know your audience. The presentation must be adjusted to appropriately address the backgrounds of the people you are speaking to. Imagine how the presentations on the same math research paper might differ for each of the following audiences:

❏ the high school math department
❏ the high school English department

❏ the parent-teacher association
❏ a contest judge who is a college math professor
❏ the students in your math class

The second key factor that affects your presentation is the amount of time you are allotted. The difference between what you can cover in fifteen minutes and what you can cover in thirty or forty minutes is astounding. You may have difficulty cutting your presentation to the allotted time. At other times, two minutes can seem like an eternity. Compressing your work into a given time slot requires that you edit the paper and look for material that can be cut or condensed. The material you can leave out and the depth of the material you present must be considered carefully. (If you are entering a contest, be sure to note all of the regulations and requirements regarding the oral presentation.)

In the planning stage, you must focus on the content of your presentation. Your audience will benefit most if you customize your presentation for them. The mathematical content of your presentation can be divided into three parts:
❏ introduction
❏ body of presentation
❏ conclusion

Let's discuss each of these parts in detail.

❏ **Introduction:** When you plan your talk, remember that you should first give your name, grade, and school and the math course you are currently enrolled in. This will orient your audience. Give the name of your paper and discuss how you found your topic. You might want to prepare a handout for the audience to engage them. This handout should not be summary information about your paper—it should be a question or challenge related to your research that will "hook" the audience. It may be your problem statement or a variation of it. It could be from another part of your paper or from a related "teaser" that isn't actually included in your paper. You might distribute the handout while you are setting up the visual aids for your presentation, before you actually start your talk. A sample handout from a paper on the arbelos follows. Notice that you don't need to know what an arbelos is to understand the handout. In this case, the handout acts like a bridge from the circumference formula, with which most people are familiar, to the paper's topic, the arbelos. (An arbelos is a plane figure defined by three semicircles arranged in a specific way.) The answer to this particular handout is addressed in the first part of the oral presentation.

PROBLEM STATEMENT: To express the area of an arbelos in terms of its axis.

INTRODUCTORY CHALLENGE: If the following semi-circles are tangent to each other with diameter lengths as shown in the diagram, find the difference between the semi-circumference of the large semi-circle and the sum of the semi-circumferences of the smaller semi-circles.

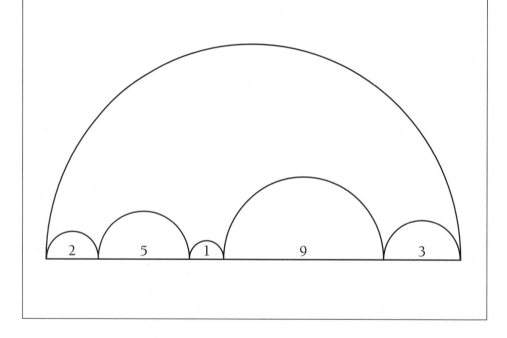

Sample handout for oral presentation. The answer is zero.

The question asked in the handout should be answered by you during your presentation. If you would like to supply your audience with a summary handout, you can distribute copies of your title and abstract at the end of your presentation. You don't want to "spill the beans" too early—you want the audience mentally at the edge of their seats waiting for your explanations of the questions you pose during the body of your talk.

❏ **Body of presentation:** The first step involved in planning the body of the presentation is to create an outline of your talk. The outline of your presentation can be written using your paper and/or the outline of your paper as a skeleton. The material will be presented in the order in which it appears in your paper. Make adjustments to meet audience and time restrictions. The outline of the body should have the following three main sections:

1. Problem Statement: You need to state your problem succinctly and give any introductory material that the audience would need in order to understand the problem statement. You might decide to include your problem statement on your initial handout.

2. Survey of Related Research: In this section, you present the major points of the article you read. You need to cover essential background material. Lead the audience through the article's major points as if you are discovering the mathematics with them as you proceed. Make conjectures as you proceed based on the information you have presented. Items developed in your paper through patterns and examples that led to conjectures and proofs should be presented that way. By not merely stating theorems, you will rouse the mathematical curiosity of your audience. Don't just recite facts—ask questions and let the audience see the material that led to the conjectures and claims you eventually made. In other parts of your talk, answer the questions you have posed.

   If there are several similar proofs in your article, don't go through all of them in detail in your oral presentation. Pick just one to explain. For example, Sara gave a presentation on her paper, which was based on the article on escribed circles in Chapter 7. The article featured the derivation of the length of the radius for one circle. Lengths for the radii of the other two circles were given but not derived. Sara derived them in her paper. Since the derivations were similar, she presented only one proof in her oral presentation.

3. Original Conjectures, Extensions, Proofs, and Findings: Much of the original work in your paper may be interspersed with your survey of the article, since you added material between the lines, tested claims in the article, and so on. This original work should not be separated from the Survey of Related Research section of your talk. Material that is an extension of the article's findings can be included after the Survey of Related Research section. For example, Jocelyn's spreadsheet in Chapter 5 represents an extension of her article. Her article focused on area, and she investigated perimeter based on the article's findings. Your original investigation can include conjectures as well as original proofs. If you read several articles and tied the results together, include this combination of results as an original finding. Your oral presentation should stress your original conjectures and discoveries.

❏ **Conclusion:** The conclusion of your talk is not a summary of your paper. You can briefly summarize the major points of your paper as an introduction to your recommendations for further research. Your presentation concludes as your paper ended, with these recommendations. Present the recommendations from your paper and explain why they are a logical next step based on the research you have completed.

The amount of time you devote to each section depends on your particular research. As the expert on your topic, you need to make decisions on what to include. If most of your research focused on the article, then your section on original findings might be very short. If you made elaborate conjectures after reading your article, a majority of your talk may deal with the original findings. The outline is subject to change based on its conformity to the time allotment. Revise the outline as necessary during the practice stage of your oral presentation development.

Visual aids and ancillary materials can be used to make significant time adjustments, whether you need to condense your presentation or have extra time to fill. Later in this chapter, you will read about coordinating your outline with your time allotment and visual aids.

Preparing Materials for Your Oral Presentation

Before you create any ancillary materials for your presentation, be sure each one has a specific purpose. Employ art only where it enhances mathematical understanding. The artistic quality of any visual aid you produce should be superseded by its effectiveness in transmitting information crucial to your report. It is hard to imagine a mathematics presentation that does not use visual aids. If a picture is worth 1000 words, then visual aids can transmit information in an efficient, time-saving manner. There are several different forms of visual aids:

❑ posters
❑ displays
❑ overhead transparencies
❑ felt boards
❑ manipulatives
❑ models
❑ computers
❑ graphics calculator overhead viewscreens
❑ slides and videotape

Choose your visual aids with discretion. Consider supplying your audience with some of these visual aids in a handout so they can follow your presentation. You might even make manipulatives and models available to your audience. We will discuss each of the visual aids listed above in detail.

❑ **Posters:** As you read through your presentation outline and your research paper, decide what portions lend themselves to a poster treatment. Sketch your posters on paper first. Proofread your posters as you create them. When you prepare your posters, make sure they can be read by everybody in the audience. If the size of the audience necessitates a large room, posters may be of little value to people in the back rows. This is usually not a problem with math contests, however. In most cases, your presentation will be in a room suitable for the use of posters.

The computer and the photocopier work well together to provide professional results. You can set up your text on the computer—including diagrams, tables, and text taken straight from your paper—enlarge them on a copy machine, and paste them up on your poster. For some applications, characters generated by a large font size on the computer may be sufficient, and you may not need to enlarge the printout. Choose a suitable font, no script, and nothing fancy. In general, use color only to help illustrate a point; do not use color for color's sake. Many of your posters may be black and white. If you are artistic, you can create most of your posters by hand. Remember, as in your paper, clarity is paramount. Keep your posters as uncluttered as possible.

Posters should be done on foam-core (a foam-filled paper "sandwich"), which is available at art and stationery stores. Oaktag and mat board are not as good, but they can be used. Oaktag and mat board do not stay flat, and this creates problems at presentations. Foam-core is perfect when you have a ledge, such as a blackboard ledge, to stand your posters on, rather than securing them with tape, because then you can use the reverse side as well during your presentation. Foam-core comes in white and colors. You can draw on them, paste things on them, and so on. Notice how neatly Robin's posters fit on a blackboard ledge as shown below.

These posters were made from paste-ups of colored paper and computer printouts and enlargements. Color is used with discretion—only where it helps convey a mathematical point; most of Robin's posters are black and white text. Using pointers prevents you from blocking the material on your posters as you explain it. A three-foot dowel, available at any lumberyard, is perfect for a pointer.

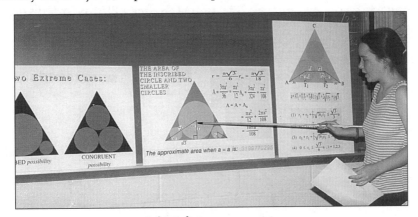

Robin's foam-core posters.

What should you put on your posters? Certainly, any diagrams you intend to refer to should be put on a poster. Re-creating them on the blackboard wastes time, and doesn't produce neat results. Geometric concepts always benefit from visual aids. Tables that display patterns should be visible to your audience. Key counterexamples to your original conjectures can be shown on a poster. Reprinting long algebraic derivations on posters allows you to use the pointer to explain each step. It is virtually impossible to understand a long algebraic derivation when it is presented orally with no visual accompaniment. As you look at Robin's posters, try to imagine how difficult it would be to *hear* this development without *seeing* it. Most likely, the speech would be verbally cumbersome, and its essence could elude the audience.

If you need to highlight portions of your poster as you speak about them, you can cover your posters with clear plastic and write on them with erasable overhead markers. The posters can be used again once you erase the markings on the clear plastic. You can even overlay several sheets of clear plastic that can be exposed one at a time to develop a concept gradually. Use your imagination and creativity to expand on the poster suggestions in this section—but be careful not to let your artistic sense detract from the mathematical message of the poster.

❏ **Displays:** For some contests, you will use a display table rather than the front of a classroom. The judge will walk right up to your table, so there is no need for posters with large type. In these cases, you can make a display. The display can be made out of foam-core. It can feature actual diagrams, tables, and text from your paper. The display should be free-standing, and it should give you as much room to display material as possible. Before you construct a display, you need to plan exactly what will be pasted onto it. Then you can take measurements and start construction.

Schematics for making one type of display out of foam-core are shown below. You will need a razor blade and a ruler. Separate the board into three folded sections. Create a "hinge" in the foam-core by scoring a line along the back of the board; do not cut all the way through the board.

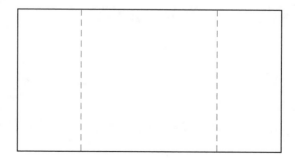

The three faces of the display.

After the lines are scored, bend the two outside faces forward to create a free-standing display, as shown.

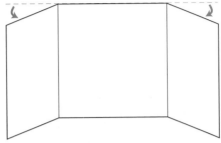

The "tri-fold" allows the display to stand on its own.

You have full use of the trapezoidal space formed by the display and the table. Cut out a piece of foam-core to fit in the space, as shown.

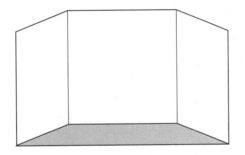

The trapezoid gives additional space.

So far, the design has four faces you can use to display material about your research. The title can appear on a fifth face, which starts with two cuts at the top of the sides.

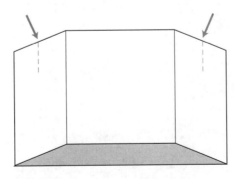

The two cuts shown will accept another face for the display.

Cut out one more piece of foam-core and make two cuts as shown in the figure below. Fit the rectangular headpiece onto the display by slipping the pieces of foam-core together at the cuts. The resulting display is sturdy as well as aesthetically pleasing.

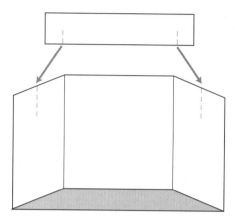

The title of your presentation goes on the headpiece.

You can paste up pages of your research paper, diagrams, enlargements, reductions, overlays, and any other visual that would enhance your presentation and can be mounted on the display. Carefully plan the allotment of space on your display. Use your imagination to extend and alter the display described above, so it meets the needs of your presentation. A completed display is shown below.

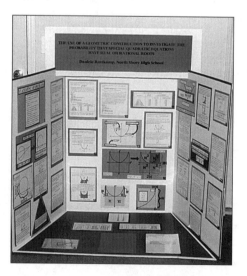

A completed display.

❏ **Overhead Transparencies:** If you are making a presentation to a large group of people who are seated too far away for a display or poster to be effective, you can rely on transparencies, an overhead projector, and a screen. Overhead projectors can enlarge transparencies so they can be read from the back of an auditorium or gymnasium. The ability to handle large rooms is not the only benefit of using transparencies, however. Transparencies are equally effective in classrooms. If you plan to use many transparencies in your presentation, keep them numbered in a three-ring binder with a plain sheet of white paper in between each transparency.

Transparencies can be handmade or produced on photocopiers or thermal-transfer machines. You can take material directly from your paper and reproduce it on a transparency. The diagrams from your paper will then appear on the transparency exactly as they appear in your paper. Once such a transparency is made, you can write on it with erasable overhead pens to highlight material during your presentation. The easiest and neatest way to display a table is to copy it onto a transparency. The table can be enlarged on a photocopier and then transferred to the transparency if necessary. The same procedure can be used for long proofs and algebraic and arithmetic derivations. Hours of work that would normally be done by hand can be saved by making transparencies in this manner. Transparencies printed in color are available at higher cost; a local print shop can quote you a price. For most applications, black print is sufficient. You can also purchase colored transparency material to use in shading and highlighting a part of a transparency. Simply place the colored acetate over the part you want to highlight. Transparencies of graph grids can be purchased, or you can reproduce a piece of graph paper onto a transparency. This allows you to graph during your presentation.

You can use permanent markers to create your own transparencies by hand. As with machine-produced transparencies, once a transparency is completed you can use erasable overhead pens to highlight material during your presentation. When you erase these markings, the work done in permanent marker will not be erased. If you make the transparencies by hand, you will be able to use permanent markers in different colors. (If you make an error using the permanent marker, try erasing it with alcohol.)

All types of transparencies can be used as part of an overlay. You can superimpose transparencies on one another to build a final picture that is developed and explained step by step. This procedure can be used for an algebraic derivation, a proof, a geometrical construction—the possibilities are limitless. When you create the overlays, you must make sure everything lines up when the transparencies are superimposed on one another. You can create them this way by tracing, binding the transparencies with a staple so they stay aligned as one unit. If you bind the finished transparencies as a "book" on the left, they can be superimposed in only one order. Another way to bind them is shown below. The four outside overlays each use a hinge of transparent tape where they meet the central image. The four transparencies can then be superimposed in any order, giving you versatility.

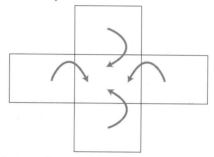

The four overlays can be superimposed in any order.

There is one disadvantage to using transparencies. With only one projector, only one image can be displayed at one time. With posters and displays, many images are in view at one time. If your presentation involves material that requires scrutiny of several images at one time, using a transparency will handicap you. Many presentations use posters *and* transparencies to combine the advantages of both.

❏ **Felt Boards:** Felt boards are useful in presentations that involve geometry. If, during the development, it is advantageous to be able to move shapes around on the plane, a felt board is perfect. A felt board is easy to make. Decide on the size you will need. You can probably find an inexpensive piece of scrap wood at a lumberyard; 1/2-inch plywood or particle board is excellent. Go to a fabric or crafts store and purchase enough felt or flannel in a solid color to completely cover the wood. You will need to add about 6 inches to each dimension. Stretch the felt over the board and staple it to the back of the board. Once you have chosen the background color, purchase contrasting colors of felt to make the shapes you need to manipulate. Trace onto the felt and cut out the shapes with scissors. Place the shapes on the felt board—notice that they "stick." During your presentation, the felt board can be placed on a blackboard ledge. To the right is a sample felt board.

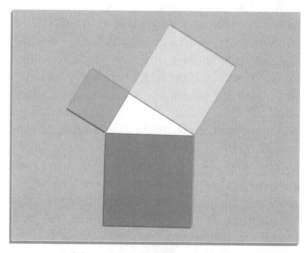

Felt boards are handy and reusable.

❏ **Manipulatives:** When you rely on the physical handling of objects to illustrate a mathematical idea, you are using some sort of manipulative. (Measuring instruments such as the protractor, compass, and ruler are not considered manipulatives.) If you ever used base-ten blocks, c-rods, pattern blocks, or algebra tiles you used a manipulative—the sight and physical manipulation of the object help teach a concept. The cutouts you make for a felt board are manipulatives. You can make manipulatives out of a wide variety of materials—clay, cardboard, velcro, elastic, transparencies, string, styrofoam, paper cups, popsicle sticks, wood, tin cans, plastic soft drink bottles, and so on. The only limit is your imagination. If you create a manipulative to enhance your mathematical presentation, you might want to allow your audience to try it if this is practical.

If your paper involves transformations, you might want to make a manipulative overlay out of a transparency and a sheet of paper. Images that need to be compared can be placed on the paper and transparency. The transparency can be superimposed over the paper and moved around to a position that helps you display your findings. For example, you can show that angles are congruent, shapes are similar, or lines are perpendicular.

Let's look at a simple manipulative that can be used to illustrate the mathematical definition of an ellipse. An ellipse is the set of points the sum of whose distances from two given, fixed points is a constant. You will need wood, two nails, string, and chalk. Drive the two nails into the board:

The two fixed points that will determine the ellipse.

Tie one length of string to both nails. The string should be longer than the distance between the two fixed points, as shown below.

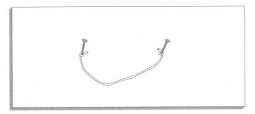

The string hangs loosely on the two nails.

Take the chalk (or any writing implement) and pull the string taut:

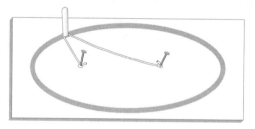

The chalk is on one point of the ellipse.

Drag the chalk around, keeping the string taut at all times. The taut string has constant length, so the sum of the two line segments is constant. As you drag the chalk, you will see the ellipse take shape, and your audience will better understand why an ellipse is more than just an "oval."

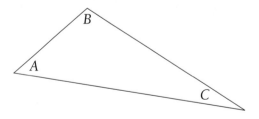

The ellipse takes shape.

Let's use a manipulative to show that the sum of the angles of a triangle is 180 degrees. You will need scissors and a straightedge. Draw any triangle. If your audience participates, each member can draw a triangle. Label the angles A, B, and C, writing the letters inside the angle as shown.

Cut out the triangle carefully. "Rip" the angles off at the lines shown below. Keep the three "angles" and discard the rest of the triangle.

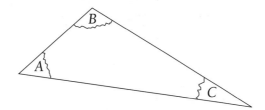

Draw a straight angle on a sheet of paper visible to your audience. Group the three ripped angles together so their vertices coincide and the angles are placed ray-to-ray, as shown.

Notice that when the angles are placed adjacent to each other their sum appears to be 180 degrees. This is not a proof, but the fact that it occurs in all of the trials by the audience leads to a conjecture and is the motivation to attempt a proof.

The next manipulative relates the Pythagorean theorem to the concept of area. Construct a right triangle on paper. Make it large, so it can be handled easily. Construct a square adjacent to each side as shown. We will label the squares with their areas.

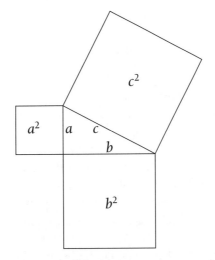

Fold the largest square along the hypotenuse so it goes under the triangle and parts of the two smaller squares. Fold the small square along the short leg, over the triangle, so it overlaps the triangle. Take scissors and cut out the pieces of a^2 and b^2 that are not overlayed on c^2. You should be cutting off the three shaded triangles shown below.

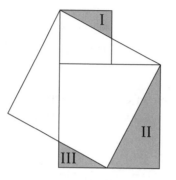

Look at the manipulative. Most of c^2 is covered by the remaining parts of a^2 and b^2. See if you can arrange the three cut-off triangles on top of c^2 to fill in the spaces not covered by a^2 and b^2. This illustrates that the sum of the areas of the two smaller squares equals the area of the large square, or $a^2 + b^2 = c^2$. It is not a proof.

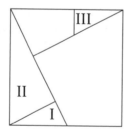

Many geometric properties lend themselves to paper folding, cutouts, and overlays. As you do your research, always think about ways manipulatives can aid your presentation as well as your own understanding.

❏ **Models:** If you are doing research on three-dimensional geometric objects, you may conduct your research and represent your findings using diagrams on paper. Perspective drawings may be sufficient for your explanations. Sometimes, however, these drawings are difficult to create, and even when they are done well they may be too confusing to illustrate a point. Let's look at two examples.

Eddie was doing research on polyhedra. He needed to research the number of faces, edges, and vertices the dodecahedron has. He had some theories, but he wanted to conclusively test his conjectures. He constructed a dodecahedron from a prefabricated manipulative kit. His work is shown on the next page.

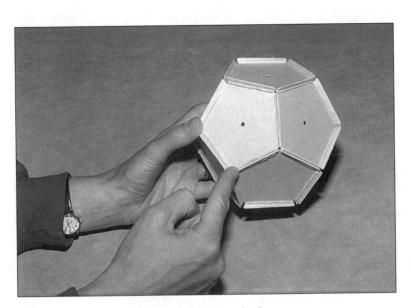

Eddie's model dodecahedron.

This manipulative allowed Eddie to prove his conjecture directly. For more intricate polyhedra, models become a necessity, especially when you are trying to convey your point in a short presentation.

Gino was doing research on applications of conic sections to satellite antennas. He needed to demonstrate how a cone and a plane intersect to form a parabola and a hyperbola. Gino stuffed a funnel with clay to form the cones he needed. He was then able to use a plastic knife to simulate the cutting of the cone by the plane. The different cross sections created by the different cuts allowed Gino to display the curves and show how they were formed. The model did a better job than the perspective drawings of the same cuts. Making a model is the three-dimensional version of the "draw a diagram" problem-solving strategy.

❏ **Computers:** Much modern mathematics research involves the use of computers. You can generate tables, search for patterns, use graphing software, or create an original program to perform a function integral to your research. The computer's ability to store and print data makes it a valuable tool for forming conjectures.

In many cases, you can print your programs and sketches and the text of your paper and reproduce them onto a transparency or poster for your presentation. Other times, it may be necessary to demonstrate the use of the computer "live" during your presentation. If your presentation is part of a contest requirement, check the rules for regulations regarding computer use. Make sure you have equipment suitable for your audience and room size.

❏ **Graphics Calculator Overhead Viewscreens:** The graphics calculator is capable of much more than drawing and analyzing graphs. If you have applied the graphics calculator to your research, you might report the procedures and findings in your paper and reproduce them on a poster or transparency. If you want to use the graphics calculator during your presentation, you will need to gain access to an overhead viewscreen that displays the calculator window onto a screen.

Decide whether you absolutely need to have the calculator at your presentation. In the past, students without access to one of these viewscreens have actually placed their calculators onto a photocopier and reproduced the image right from the calculator's window. This may not give satisfactory results on all copy machines. You can always copy the display by hand or by drawing on a computer. Packages are available for some graphics calculators that include hardware and software to link the calculator to a computer. The calculator's display can then be reproduced on the computer screen and printed. These graphics can be merged into your paper and used to make transparencies.

❏ **Slides and Videotape:** Slides and videotape can be used if you have the necessary equipment available to you. Slides can be taken with a conventional camera and projected onto a large screen. You can write a script and record a narration on an audiocassette for use with the slides. If your presentation is part of a contest requirement, make sure the use of slides and audiotape is permissible. Slide use should be reserved for displaying photographs and other materials that cannot be reproduced onto a transparency. Eddie's research on polyhedra would lend itself well to slides of the models he built. Slides taken from photographs of models would allow him to display his models without building and/or transporting them.

Video cameras can aid presentations on research topics involving motion. For example, the presentation of a paper on rates could make use of a video camera.

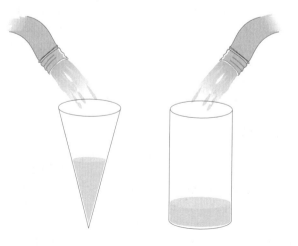

An inverted conical reservoir is being filled with water at the rate of 3 pints per minute. As time passes, the water level rises, but it rises at a decreasing rate. (If the water were poured into a right circular cylinder, the water level would rise at a constant rate.) A video of these two containers being filled at equal, constant rates would illustrate the differences in the water levels at any given time and would show the decreasing rate of the cone's water level.

In your math classes you may have studied about transformations in the plane. These include reflections, rotations, translations, and dilations. A presentation on plane transformations would make extensive use of transparencies, posters, manipulatives, and possibly a felt board. If your research involved transformations in three dimensions, you could build models and videotape a demonstration of some of the transformations.

Practicing Your Oral Presentation

After you have planned your presentation and prepared your materials, it is time to practice your presentation. You need to coordinate your outline, the time you will spend on each item, and the materials needed for each item. Create a Coordination of Presentation sheet to organize everything. When it has been completed, tested, and revised, print this sheet in large type on a large sheet of paper. It can be placed on the desk during your presentation, next to a watch or clock. At a glance, you will be able to see a schematic of your entire presentation. Table 9.1 shows a sample Coordination of Presentation sheet. The numbers under "Outline" refer to your presentation outline, which should appear in large type next to this coordination sheet. "P" refers to poster, "T" to transparency, "M" to model, and "D" to display.

Table 9.1 Sample Coordination of Presentation Sheet for a Twenty-Minute Presentation

Outline	Pages in Paper	Materials Needed	Time
I.A.	1–3	P1, P2, T1–T3	2 min.
I.B.	3–7, 11	P3, T4	4 min.
II.A.	8–14	T5, T6, D1	8 min.
II.B.	15–18	M1, T7	11 min.
II.C.	19–22	P4, P5	14 min.
III.A.	22–24	blackboard	16 min.
III.B.	25	P6	17 min.
V.A.	26–29	T8, T9	20 min.

Notice that "Time" is kept as a running total so you can compare your pace to your timepiece. A stopwatch works ideally with this column. Once you have completed the first draft of your coordination sheet, you can start rehearsing. As you rehearse, have a copy of your paper with you with pieces of stick-on notepaper as bookmarks that can you can use to quickly find specific parts of your paper. These stick-on notes can be coordinated with the information on the Coordination of Presentation sheet. Changes will be made in the coordination sheet during the course of your rehearsals. Do not make the final, large-type copy of the sheet and your outline until you have worked out all of the "kinks" in your talk.

How can you make your presentation smooth, and reduce nervousness? Know your audience. The presentation must be geared to their level of mathematics and their familiarity with your research. (Possibly your paper was submitted weeks in advance and the judges have read it.) Practice often, without an audience in the beginning. Get a video camera and ask someone to tape your rehearsals, or set up a camera on a tripod and tape them yourself. Watch the tapes and critique them. Athletes, politicians, teachers, and actors do this, and you, too, can use tapes to improve your delivery. Audiotape your rehearsals. Listen to the audio without watching the video. Is your communication precise, correct, and mathematically sophisticated? Are you speaking in full, well-defined sentences?

Once you are well rehearsed, begin "dress rehearsals" with audiences. Parents, friends, and siblings may not understand the research itself, but they will help you simulate the conditions of having to perform in front of a live audience. If you plan to enter a judged contest, rehearse by having your teacher watch your presentation and make a list of positive points and aspects that need work. Make presentations to your math class, the math department, math club, school faculty meeting, and so on. Always remain receptive to constructive criticism. Field questions from your practice audiences. Try to anticipate questions the judges might ask, and prepare the answers. Solid preparation is your best defense against nervousness. Show enthusiasm, speak clearly, and pace yourself. Remember to pause, and don't rush. You should be very confident once you have ironed out all the wrinkles in your presentation.

Many schools require students to give an exhibition as a condition of graduation. This exhibition can take many forms. Consult your teacher and guidance counselor to find out whether your research and presentation in mathematics can qualify as your exhibition. As you review the steps you've followed over the chapters of this book, note carefully ways in which your research satisfies your school's requirements for performance assessments and exhibitions. Good luck!

Appendix A
Resources

In Chapter 6, you read about finding a topic. There are many resources you can use to find a topic or to locate additional material for a topic you have already started. This appendix lists dozens of possible resources under the following categories:
- ❏ periodicals
- ❏ problem solving
- ❏ research topics

These categories are not mutually exclusive. For example, a book on problem solving may inspire a research topic or provide an extension to your research topic. All books and articles in all the categories are potential sources of material for research papers. Standard high school textbooks are not listed here. In cases where you need supplementary material from a textbook, see your teacher or the math department chairperson. The periodicals section also includes journal articles that students have used in their research.

Periodicals

In Chapter 6, you learned how to use journal articles to find a topic. Below is a list of popular journals. You may find some of these publications cited as you create your bibliography tree. Look in the periodicals section of your high school library, local college library, or public library. If you are interested in a particular back issue or subscription information or have other questions, use the addresses given below to write to the publisher.

College Mathematics Journal
Mathematics Association of America
1529 Eighteenth Street, NW
Washington, DC 20036-1385

Consortium
UMAP Catalog and
HiMAP Catalog
COMAP
60 Lowell Street
Arlington, MA 02174-1295

Duodecimal Bulletin
The Dozenal Society of America
c/o Math Department
Nassau Community College
Garden City, NY 11530

Games Magazine
19 West 21st Street
New York, NY 10010

Mathematics and Informatics Quarterly
Department of Mathematics—Box 21
Rose-Hulman Institute of Technology
Terre Haute, IN 47803-3999

Mathematics Magazine and
The American Mathematical Monthly
Mathematical Association of America
1529 Eighteenth Street, NW
Washington, DC 20036

The Mathematics Teacher
National Council of Teachers of Mathematics
1906 Association Drive
Reston, VA 22091

Mathematics Teaching
Association of Teachers of Mathematics
7 Shaftsbury Street
Derby, Great Britain DE3 8YB

Quantum
National Science Teachers Association
1840 Wilson Boulevard
Arlington, VA 22201-3000

School Science and Mathematics
Curriculum and Foundations
Bloomsburg University
400 East Second Street
Bloomsburg, PA 17815-1301

Scientific American
415 Madison Avenue
New York, NY 10017-1111

Some periodicals have an annual index that you might find useful. Ask the librarian if you need help tracking down a journal or a specific article. If you plan on writing to the publisher for information, make sure you send your letter immediately. Call to find out exactly to whom your letter should be addressed. You must allow time for a response, and you don't want to delay your research work unnecessarily. You might be able to secure the information by telephone.

Some Selected Journal Articles

The following is a list of articles on which students have based research papers. The articles are from *Mathematics Teacher*, *Mathematics and Informatics Quarterly*, and *High School Mathematics and Its Applications Project*.

Mathematics Teacher

March 1966	Pseudo-Ternary Arithmetic
April 1966	A Geometric Approach to the Conic Sections
May 1966	Even More on Pascal's Triangle and Powers of 11
October 1966	Radii of the Apollonius Contact Circles
November 1966	Geometric Solution of a Quadratic Equation
February 1967	Auxiliary Lines and Ratios
May 1967	Five-Con Triangles
November 1967	Investigation to Discovery with a Negative Base
April 1968	Divisibility by 7, 11, 13 and Greater Primes
January 1969	Why Not Relate Conic Sections to the Cone?
February 1969	Exploring Geometric Maxima and Minima
May 1970	An Interesting Relationship Among the Roots of a Cubic Equation
May 1971	Circular Coordinates: A Strange New System of Coordinates
January 1974	All Three-Digit Integers Lead to . . .
May 1974	Area Ratios in Convex Polygons
February 1975	Serendipitous Discovery of Pascal's Triangle

May 1994   Nested Platonic Solids

Mathematics and Informatics Quarterly

February 1991   The Bobillier Theorem
February 1991   Simple Properties of the Orthodiagonal Quadrilaterals
February 1991   What Is the Use of the Last Digit?
August 1991   The Folded Square
December 1991   Factorization of Quadratic Polynomials in Two Variables
March 1992   Figures of Equal Area
March 1992   On the Bobillier Theorem
March 1992   Orthogonal Quadrilaterals Again
May 1992   On the Malfatti Problem for the Equilateral Triangle
September 1992   The Median Toward the Hypotenuse
November 1992   Cyclic Quadrilaterals with Integral Sides and Diagonals
May 1993   Tangent Nine-Point Circles
November 1994   A Telephone Dialing Question
November 1994   An Etude on Bisectors
November 1994   Equations, Inequalities and Graphs
March 1994   Expected and Unexpected Solutions
June 1994   Algebraic Relations for the Inscribed Orthodiagonal Quadrilaterals

High School Mathematics and its Applications Project (HiMAP)

The following "modules" are sold separately. You can write to COMAP to order a catalog of all modules offered. The address is given on page 134.

A Mathematical Look at the Calendar
A Uniform Approach to Rate and Ratio Problems
Applications of Geometrical Probability
Architecture Designpack
Businesspack
Codes Galore
Decision Making and Math Models
Drawing Pictures with One Line
Enviropack
Loads of Codes
Medipack
Sociopoliticopack I, II
The Mathematical Theory of Elections

Problem Solving

Most books on problem solving are timeless. Some of the best classic problems are years, decades, even centuries old. Libraries and bookstores have many recently published titles about problem solving, but don't shy away from a book because it is old. Many problem-solving books from previous decades are still available in bookstores today or can be found in a library. Be aware that some older books may be more difficult to find.

Many problem-solving books give solutions and hints for some or all of the problems. You can use the problems to improve your problem-solving skills, to enhance your research paper, to find a topic, or for just plain fun. Use the following list, or find other books in your library. Although the following book list, in totality, contains over 1000 brainteasers, it merely scratches the surface of the available books on problem solving.

Artino, R., Gaglione, A. and Shell, N. *The Contest Problem Book IV: American High School Mathematics Examination 1973–1982.* Washington, D.C.: Mathematical Association of America, 1983.

Barr, S. *Mathematical Brain Benders.* New York: Dover, 1982.

Bates, N., and Smith, S. *101 Puzzle Problems.* Concord, Mass.: Bates, 1980.

Bolt, B. *The Amazing Mathematical Amusement Arcade.* New York: Cambridge University Press, 1984.

Butts, T. *Problem Solving in Mathematics.* Glenview, Ill.: Scott, Foresman, 1973.

Conrad, S., and Flegler, D. *The First High School Math League Problem Book.* Tenafly, N.J.: Math League Press, 1989.

Conrad, S., and Flegler, D. *The Second High School Math League Problem Book.* Tenafly, N.J.: Math League Press, 1992.

Ecker, M. *Getting Started in Problem Solving and Math Contests.* New York: Franklin-Watts, 1987.

Flener, F. *Mathematics Contests: A Guide for Involving Students and Schools.* Reston, Va.: NCTM, 1990.

Frohlichstein, J. *Mathematical Fun, Games and Puzzles.* New York: Dover, 1967.

Gardner, M. *Perplexing Puzzles and Tantalizing Teasers.* New York: Dover, 1969.

Gardner, M. *Wheels, Life and Other Mathematical Amusements.* New York: W. H. Freeman, 1983.

Gilbert, G., Krusemeyer, M., and Larson, L. *The Wohascum County Problem Book.* Washington, D.C.: Mathematical Association of America, 1993.

Greenes, C., Schulman, L., Spungin, R., Chapin, S., and Findell, C. *Mathletics—Gold Medal Problems.* Providence, R.I.: Janson Publications, 1990.

Greitzer, S. *International Math Olympiads 1959–1977.* Washington, D.C.: Mathematical Association of America, 1978.

Grosswirth, M., and Salny, A. *The Mensa Genius Quiz Book.* Reading, Mass.: Addison-Wesley, 1981.

Grosswirth, M., and Salny, A. *The Mensa Genius Quiz Book 2.* Reading, Mass.: Addison-Wesley, 1983.

Halmos, P. *Problems for Mathematicians Young and Old.* Washington, D.C.: Mathematical Association of America, 1991.

Herr, T., and Johnson, K. *Problem Solving Strategies—Crossing the River With Dogs and Other Mathematical Adventures.* Berkeley, Calif.: Key Curriculum Press, 1994.

Houghton, G. *Common Sense Puzzles.* New York: Hart, 1984.

Hunter, J. *Challenging Mathematical Teasers.* New York: Dover, 1980.

Klamkin, M. *International Math Olympiads 1978–1985.* Washington, D.C.: Mathematical Association of America, 1986.

Mira, J. *Mathematical Teasers.* New York: Barnes & Noble, 1970.

Mosteller, F. *Fifty Challenging Problems in Probability.* Reading, Mass.: Addison-Wesley, 1965.

Phillips, H. *My Best Puzzles in Mathematics.* New York: Dover, 1961.

Reeves, C. *Problem Solving Techniques Helpful in Mathematics and Science.* Reston, Va.: NCTM, 1987.

Salkind, C. *The Contest Problem Book I: American High School Mathematics Examination 1950–1960.* Washington, D.C.: Mathematical Association of America, 1961.

Salkind, C. *The Contest Problem Book II: American High School Mathematics Examination 1961–1965.* Washington, D.C.: Mathematical Association of America, 1966.

Salkind, C. *The Contest Problem Book III: American High School Mathematics Examination 1966–1972.* Washington, D.C.: Mathematical Association of America, 1973.

Serebriakoff, V. *Puzzles, Problems, and Pastimes for the Superintelligent.* Englewood Cliffs, N.J.: Prentice-Hall, 1983.

Shushan, R. *Games Magazine Big Book of Games.* New York: Workman, 1984.

Yawin, R. *Math Games and Number Tricks.* Middletown, Conn.: Field Publications, 1987.

Research Topics

In Chapter 6, you learned that articles from periodicals and books can help you find a research topic. Some books have short mathematical "essays" that are excellent springboards for research papers. These essays are just like journal articles. A single book might contain dozens of essays. Some of these books are listed below. Your library will have other titles; also check with the math department chairperson and your teacher to see if they have some of these books or others like them.

Aaboe, A. *Episodes from the Early History of Mathematics.* Washington, D.C.: Mathematical Association of America, 1964.

Adler, I. *Readings in Mathematics.* Lexington, Mass.: Ginn, 1972.

Austin, J. *Applications of Secondary School Mathematics—Readings from the Mathematics Teacher.* Reston, Va: NCTM, 1991.

Beckenbach, E. *An Introduction to Inequalities.* Washington, D.C.: Mathematical Association of America, 1961.

Beiler, A. *Recreations in the Theory of Numbers.* New York: Dover, 1966.

Bolt, B. *More Mathematical Acvtivities.* Cambridge: Cambridge University Press, 1988.

Bowers, J., and Bowers, H. *Arithmetical Excursions.* New York: Dover, 1961.

Chinn, W. *First Concepts of Topology.* Washington, D.C.: Mathematical Association of America, 1966.

Consortium for Mathematics and Its Applications. *High School Lessons in Mathematical Applications.* Lexington, Mass.: COMAP, 1993.

Coxeter, H., and Greitzer, S. *Geometry Revisited.* Washington, D.C.: Mathematical Association of America, 1967.

Croft, H., Falconer, K., and Guy, R. *Unsolved Problems in Geometry.* New York: Springer-Verlag, 1991.

Dalton, L., and Snyder, H. *Topics for Mathematics Clubs.* Reston, Va.: NCTM, 1990.

Davis, P. *The Lore of Large Numbers*. Washington, D.C.: Mathematical Association of America, 1961.

Department of Mathematics and Computer Science, North Carolina School of Science and Mathematics. *New Topics for Secondary School Mathematics: Geometric Probability*. Reston, Va.: NCTM, 1988.

Department of Mathematics and Computer Science, North Carolina School of Science and Mathematics. *New Topics for Secondary School Mathematics: Matrices*. Reston, Va.: NCTM, 1988.

Eccles, F. *An Introduction to Transformational Geometry*. Menlo Park, Calif.: Addison-Wesley, 1971.

Gardner, M. *Mathematical Circus*. New York: Alfred A. Knopf, 1979.

Gardner, M. *Mathematical Magic Show*. Washington, D.C.: Mathematical Association of America, 1990.

Garland, T. *Fascinating Fibonaccis: Mystery and Magic in Numbers*. Palo Alto, Calif.: Dale Seymour, 1987.

Grossman, I. *Groups and Their Graphs*. Washington, D.C.: Mathematical Association of America, 1964.

Guy, R. *Unsolved Problems in Number Theory*. New York: Springer-Verlag, 1991.

Henle, J. *Numerous Numerals*. Reston, Va.: NCTM, 1975.

Hofstadter, D. *Godel, Escher, Bach: An Eternal Golden Braid*. New York: Vintage, 1980.

Honsberger, R. *Ingenuity in Mathematics*. Washington, D.C.: Mathematical Association of America, 1970.

Honsberger, R. *Mathematical Gems I*. Washington, D.C.: Mathematical Association of America, 1973.

Honsberger, R. *Mathematical Gems II*. Washington, D.C.: Mathematical Association of America, 1976.

Honsberger, R. *Mathematical Gems III*. Washington, D.C.: Mathematical Association of America, 1985.

Honsberger, R. *Mathematical Morsels*. Washington, D.C.: Mathematical Association of America, 1976.

Honsberger, R. *Mathematical Plums.* Washington, D.C.: Mathematical Association of America, 1985.

Hopkins, N., Wayne, J., & Hudson, J. *The Numbers You Need.* Detroit, Mich.: Gale Research, 1992.

Huntley, H. *The Divine Proportion: A Study in Mathematical Beauty.* New York: Dover, 1970.

Kline, M. *Mathematics in Western Culture.* New York: Oxford University Press, 1953.

Loomis, E. *The Pythagorean Proposition.* Reston, Va.: NCTM, 1968.

Mottershead, L. *Metamorphosis—A Source Book of Mathematical Discovery.* Palo Alto, Calif.: Dale Seymour, 1977.

Niven, I. *Mathematics of Choice—How to Count Without Counting.* Washington, D.C.: Mathematical Association of America, 1965.

Niven, I. *Maxima and Minima Without Calculus.* Washington, D.C.: Mathematical Association of America, 1981.

Niven, I. *Numbers: Rational and Irrational.* Washington, D.C.: Mathematical Association of America, 1961.

Olds, C. *Continued Fractions.* Washington, D.C.: Mathematical Association of America, 1963.

Ore, O. *Graphs and Their Uses.* Washington, D.C.: Mathematical Association of America, 1963.

Ore, O. *Invitation to Number Theory.* Washington, D.C.: Mathematical Association of America, 1967.

Packel, E. *The Mathematics of Games and Gambling.* Washington, D.C.: Mathematical Association of America, 1981.

Petersen, I. *The Mathematical Tourist—Snapshots of Modern Mathematics.* New York: W. H. Freeman, 1988.

Posamentier, A., and Salkind, C. *Challenging Problems in Algebra 1.* New York: MacMillan, 1970.

Posamentier, A., and Salkind, C. *Challenging Problems in Algebra 2.* New York: Macmillan, 1970.

Posamentier, A., and Salkind, C. *Challenging Problems in Geometry 1*. New York: Macmillan, 1970.

Posamentier, A., and Salkind, C. *Challenging Problems in Geometry 2*. New York: Macmillan, 1970.

Ribenboim, P. *The Book of Prime Number Records*. New York: Springer-Verlag, 1989.

Rouse Ball, W., and Coxeter, H. *Mathematical Recreations and Essays*. New York: Dover, 1987.

Rucker, R. *Mind Tools—The Five Levels of Mathematical Beauty*. Boston, Mass.: Houghton-Mifflin, 1987.

Schuh, F. *The Master Book of Mathematical Recreations*. New York: Dover, 1968.

Sinkov, A. *Elementary Cryptanalysis: A Mathematical Approach*. Washington, D.C.: Mathematical Association of America, 1966.

Sobel, M. *Readings for Enrichment in Secondary School Mathematics*. Reston, Va.: NCTM, 1988.

Steen, L. *Mathematics Today—Twelve Informal Essays*. New York: Springer-Verlag, 1984.

Stepelman, J. *Milestones in Geometry*. New York: Macmillan, 1970.

Stewart, I. *The Problems of Mathematics*. Oxford: Oxford University Press, 1987.

Stwertka, A. *Recent Revolutions in Mathematics*. New York: Franklin-Watts, 1987.

Wills, III, H. *Leonardo's Dessert: No Pi*. Reston, Va.: NCTM, 1985.

Wisner, R. *A Panorama of Numbers*. Glenview, Ill.: Scott-Foresman, 1970.

Yates, R. *The Trisection Problem*. Reston, Va.: NCTM, 1971.

Zippin, L. *Uses of Infinity*. Washington, D.C.: Mathematical Association of America, 1962.

Appendix B
Sample Pages from Students' Research Papers

Appendix B is comprised of examples of actual student work. Glance through these pages to get ideas about how pages from your paper should look. Because these are just excerpts and not complete papers, you should concentrate on their form and not on the specific mathematics.

B_1: Sample cover page of research paper.

B_2: Sample Abstract
Robin Hadley, Grade 11, "A Solution to the Degenerate Malfatti Problem in Two Dimensions for the Equilateral Triangle"

B_3: Sample Problem Statement
Daniele Rottkamp, Grade 11, "The Probability That A Quadratic Equation With Randomly-Selected Coefficients Has Real Roots"

B_4: Sample Graduated Figures
Jeff Chew, Grade 11, "Determining Conditions Under Which Triangles Can Be Five-Con Capable"

B_5: Sample Centered Algebra
Dalita Balassanian, Grade 10, "An Investigation of the Properties of the Median Drawn to the Hypotenuse of A Right Triangle"

B_6: Sample Construction
(Note that construction lines are not erased.)
Nancy Friedlander, Grade 10, "An Analysis of the Properties of Two Triangles that Have Five Congruent Parts"

B_7: Sample Diagrams
Jeff Chew, Grade 12, "Approximating the Areas of Irregular Plane Figures"

B_8: Sample Computer Program and Table
Daniele Rottkamp, Grade 11, "The Probability That A Quadratic Equation With Randomly-Selected Coefficients Has Real Roots"

B_9: Sample Recommendations for Further Research
Alexis Soterakis, Grade 11, "Necessary and Sufficient Conditions for the Existence of Cyclic Quadrilaterals with Integer-Length Sides and Diagonals"

B_{10}: Sample Page from a Math Annotation Project
Tim Bramfeld, Grade 11, "Related Rates—A Math Annotation Project"

Presented at the Dutchess County Mathematics Fair, April 1996

THE PROBABILITY THAT A CUBIC EQUATION WITH
RATIONAL COEFFICIENTS HAS EXACTLY
THREE RATIONAL ROOTS

Anthony Mackin
Grade 11
North Side High School
405 Knob Hill Drive
Smithtown, VT

ABSTRACT

        The degenerate form of the Original Malfatti Problem which applies the problem to triangles and circles in two dimensional space is stated: *How are three circles located inside a given triangles, each pair of which have an empty intersection, such that the sum of their three areas is maximal?* This problem is applied to the special case of the the equilateral triangle. Formulae for the areas of two possibilities for the maximal triple, the three greatest congruent circles and the inscribed circle with two smaller congruent circles are derived. A proof concerning the location of maximized circles is explained and their radii are discussed. The final solution is provided and interpreted. In addition to this, several suggestions are made concerning further research involving the Malfatti problem.

STATEMENT OF PROBLEM

The standard quadratic equation, in the form of $ax^2 + bx + c = 0$, gives roots of a parabola with the equation $y = ax^2 + bx + c$. Several questions arise when searching for properties of quadratic equations and the nature of their roots:

- How can we *geometrically* find the roots of a quadratic equation using circles?

- What is the probability that a quadratic equation $x^2 + bx + c = 0$, where b and c are random real numbers, has real roots?

- Do quadratic equations exist with consecutive-integer coefficients and integer roots?

- Do quadratic equations exist with consecutive-integer coefficients and rational roots?

- What is the probability that a randomly-selected consecutive-integer-coefficient quadratic equation has rational roots?

The mathematical connections between the answers to the above problems will help us discover the frequency of rational roots that are generated from consecutive-integer coefficient quadratic equations.

RELATED RESEARCH

<u>Analytic Solutions To Quadratic Equations</u>

The roots of any given quadratic equation and information about the nature of these roots can be found and solved for geometrically by using circles and geometric theorems. Let us assume that two roots of a standard quadratic equation R_1 and R_2 are given such that they are two distinct points along the horizontal axis of a Cartesian graph, as shown in Figure 1:

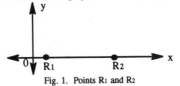

Fig. 1. Points R₁ and R₂

page 6

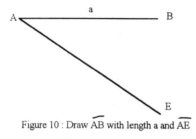

Figure 10 : Draw $\overset{\frown}{AB}$ with length a and $\overline{AE}$

Using a compass, we construct three congruent length on $\overline{AE}$:

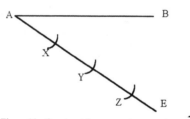

Figure 11 : Construct 3 congruent segments on $\overline{AE}$

Now, construct line segment BZ:

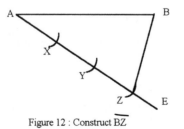

Figure 12 : Construct $\overline{BZ}$

We now copy $\angle AZB$ to Y and create a line parallel to BZ through Y:

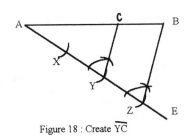

Figure 18 : Create $\overline{YC}$

The length of AC would be $\frac{2}{3}a$. Let ak be the length of the longest side of the smaller triangle, since the two triangles are similar:

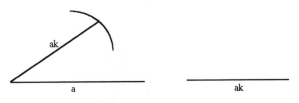

Figure 14 : The longest sides of the similar triangles

We will also let ak be the second longest side of the larger triangle:

Figure 15 : ak is the second longest side of the larger triangle

Let ak^2 be the length of the second longest side of the smaller triangle. To find ak^2, we build the $\frac{2}{3}$ construction on ak:

right triangles where the length of the altitude is the mean proportional between the lengths of the segments formed on the hypotenuse, we derive:

$$\frac{a+p}{h} = \frac{h}{a-p}$$

$$h^2 = a^2 - p^2$$

$$h^2 + p^2 = a^2 \Rightarrow a^2 = c^2 \text{ (since } p^2 + h^2 = c^2 \text{)} \Rightarrow a = c.$$

Therefore, a=c, and the measure of the median of a triangle drawn to the hypotenuse does equal half the measure of the opposite side. This proves the MH-Proposition. Next, the converse of the MH-Proposition should also be examined. (<u>NOTE</u>: throughout this investigation of the MH-Proposition, the abbreviation, "MH" will be used for the words, MH-Proposition.) Now let's prove the converse of the MH-Proposition.

If the measure of a median of a triangle equals half the measure of the opposite side, then

the triangle is a right triangle.

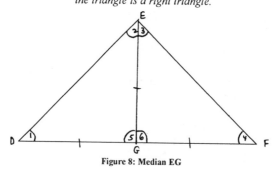

Figure 8: Median EG

By observing the labeled triangle in Figure 8, we derive that $\angle 1 = \angle 2$ (since EG=DG) and $\angle 3 = \angle 4$ (since GE=GF). We also know that $\angle 3 + \angle 4 = \angle 5$ because the sum of the measures of the two non-adjacent interior angles equals the measure of the exterior angle.

We can now continue to draw the sides of 5-Con triangles on the X-Y axes by

extending the picture shown in Figure 19 in Figure 20.

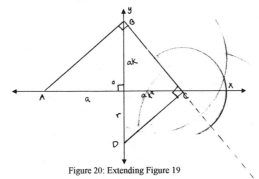

Figure 20: Extending Figure 19

$$\frac{ak}{ak^2} = \frac{ak^2}{r}$$
$$rak = a^2k^4$$
$$r = ak^3.$$

So, now the picture looks like this as shown in Figure 21:

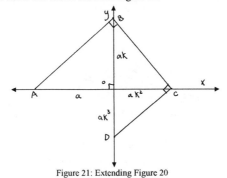

Figure 21: Extending Figure 20

Page 12

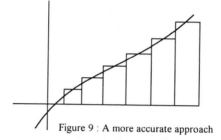

Figure 9 : A more accurate approach

Using this idea, it is possible to divide an irregular shape into sectors. The program will find two pseudoradii. Then, it will find the *average* of the two radii, then find the area of a sector of a circle with a radius of that length:

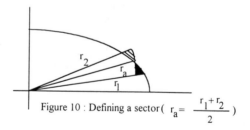

Figure 10 : Defining a sector ($r_a = \dfrac{r_1 + r_2}{2}$)

Part of the area of the sector with the striped pattern is outside of the ellipse. However, some of the sector does not overlap at the shaded pattern of the ellipse. The closer the two pseudoradii are, the more accurate the area of the sector will be to the sector of the ellipse.

The computer program below is a program which would find the areas of many sectors around the ellipse, therefore finding an approximate area:

8

real roots. Such a computer program will generate random real values for b and c, and solve for

the value of the radius length and the y-coordinate of center E. The length of a radii of circle E can

be found by using the distance formula:

$$r = \sqrt{\frac{b^2 + c^2 - 2ac + a^2}{4a^2}}$$.

The first program that is used limits the random values of b and c between the intervals of -1 to 1.

Each successive program increases the intervals of b and c and the value of N (the number of times

the program computes the real and imaginary roots) is always 10,000. The program also prints the

number of real and imaginary roots that were produced from the 10,000 trials. The following is a

sample program:

```
RANDOMIZE
FOR N = 1 TO 10,000
   LET A = 1
   LET B = 2 * RND - 1
   LET C = 2 * RND - 1
   LET RAD = SQR (((B^2) + (C^2) - 2*A*C + (A^2)) / (4*(A^2)))
   IF RAD >= ((A+C) / (2*A)) THEN LET POS = POS + 1
NEXT N
LET NEG = 10,000 - POS
PRINT POS; "WERE POSITIVE (REAL ROOTS)"
PRINT NEG; "WERE NEGATIVE (IMAG ROOTS)"
END
```

Table 1 summarizes the results from each run:

Table 1: Probabilities Over Different Intervals

Interval for b,c	Value of N (# of Trials)	Percent of Real Roots (rounded to nearest tenth of a percent)
(-1 to 1)	10,000	53.8%
(-10 to 10)	10,000	78.4%
(-100 to 100)	10,000	93.4%
(-1,000 to 1,000)	10,000	97.8%
(-10,000 to 10,000)	10,000	99.3%
(-100,000 to 100,000)	10,000	99.8%
(-1,000,000 to 1,000,000)	10,000	100%

We can see from Table 1 that as we increase the range of values for b and c, the percent of real

roots that are generated nears 100%. The last interval, from -1,000,000 to 1,000,000, is the

The case is the same for the second triple generated,

$$hk + li = 63 \qquad ik - hl = 16$$

If have a common factor of hl, ki, li, kh, exists, then the resulting numbers would also have a common factor.

RECOMMENDATIONS FOR FURTHER RESEARCH

Many questions asked during this research that remain unanswered and merit investigation, such as:

• If two primitive triples are dilated will the resulting triples from the missing sides always be primitive?

• Can incirclable quadrilaterals also be generated with integral sides and diagonals?

• Is there a cyclic and incirclable quadrilateral with integral sides and diagonals?

• If two different primitive triples with a common hypotenuse (instead of two "dilated" triples) were used to form a cyclic Quadrilateral, would the "missing" side or diagonal be integral?

The answers to these questions would enhance our understanding of cyclic and incirclable quadrilaterals.

page 1

A calculator company is coming out with a new graphing calculator model. However, the company doesn't know how many calculators to produce, what price to set the calculators at, or how much they should advertise their new product in order to get the maximum profit. There are many things to consider. For instance, if the price was set too high, the public interest in the calculators would drop. If the manufacturer produces too little of the product, it would cost more for the materials since it would cost more if you don't buy the materials by the bulk. Let's say that the product is advertised too much, the price was set too high, and the company built too many calculators. The result might be that the company spent too much money on advertising and producing too many calculators, and the calculators aren't selling due to the high price of the calculator.

The calculator company must consider all of these factors and find a way to get the maximum profit. The way to approach this is to create a function with all of these factors as variables. Then, the maximum amount of profit would be the point where the height of this function is the greatest. This is called the *maximum* of the function. Like the maximum, it is also possible to find the point of a function where the height is the least. This point is called the *minimum*. Maxima and minima can also be called *extrema*. There can be more than one maximum or one minimum:

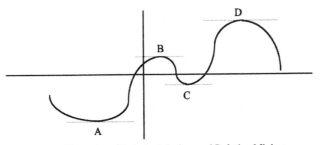

Figure 1 : Examples of Relative Maxima and Relative Minima

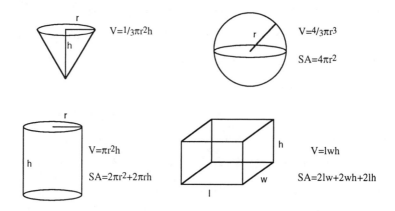

SOLVING RELATED RATES PROBLEMS: THE 10 DEMANDMENTS

Let's look at the sample problem given earlier:

A ladder 25' long leans against a wall. If the lower end is pulled away from the wall at a rate of 4 ft/sec, what is the rate of descent of the top of the ladder when the top is 7' from the floor? (Assume the wall and the floor are perpendicular to each other)

The first thing we must do is READ the entire problem. This is the first demandment in dealing with related rates problems. After we've read the problem, we should UNDERLINE key terms: the second demandment. In this case, the terms '4 ft/sec,' '7' from the floor,' and 'rate of descent' are the key terms. The third demandment is to DIAGRAM the problem, with a clock, if time is involved in the key terms. Since ft/<u>sec</u> is a component of one of our major terms, time is indeed involved, and a clock must be included. The diagram should look like Figure 5.

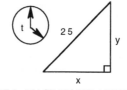

FIGURE 5: DIAGRAM OF LADDER PROBLEM

Appendix C
A Guide for Teachers and Administrators

Math Research at the High School Level

How would your students describe mathematics and their mathematics education? The typical high school mathematics program covers, to some degree, units in algebra, geometry, trigonometry, probability, statistics, logic, graphing, precalculus, and calculus. This broad range of content may not give you the chance to explore any single topic in depth. If these units are not integrated, students will not get a feel for how different branches of mathematics can team up to help them solve a problem. Understandably, many people think that all of mathematics is "already discovered." The high school curriculum traditionally presents material in succinct, forty-minute packages that often do not reflect the way math evolved. The lessons are calculated, well prepared, and timed; the students know there will be an outcome by the end of the period. As a result, mathematical concepts may seem segregated, lifeless, and predetermined.

High school programs equip students with the tools (skills) necessary to do mathematics and some applications. Due to the content requirements and time constraints of the core curriculum, there may not be enough time devoted to allowing students the oppportunity to discover, conjecture, and "play with" mathematics. Too often, "getting the answer" is the goal. This trend can suppress a student's natural curiosity. A math research program, based on the spirit of investigation and discovery that is the cornerstone of mathematics, can rekindle student enthusiasm.

If a math paper is assigned as a requirement in a core-curriculum math course, it is often completed in full outside of class without input from the teacher and other students. Why are so many students asked to write math papers with absolutely no preparation? Why do some classes have a required math paper when no training toward completing this requirement is given? Why are teachers surprised when an assigned math paper becomes a math "book report" and a questionable learning experience? Is assigning a term paper tantamount to teaching research?

> The formal papers were another matter, however. I dreaded sitting down to read and grade the stacks of barely disguised, reconstructed encyclopedia essays. . . . The best solution, it seemed to me at the time, was to abandon the term paper (Countryman, 1992, p. 66).

Mathematics research papers address many of the recommendations of the NCTM and other professional organizations, so abandoning the paper is not the solution to the problem addressed above. *Teaching appropriate skills* is the answer. Research skills are not innate; they are sophisticated, learned skills. They need to be taught, discussed, and practiced. Addressing math research formally is the first step in instilling good reading, writing, and research skills in your students. Avoiding the feelings of despair commonly experienced by teachers who assign papers requires a conscious attempt to teach about research. The purpose of this book is to introduce students and teachers to a methodical way of learning how to read mathematics, write mathematics, and do math research.

Using the Teacher's Guide

Starting a mathematics research program is an exciting undertaking. You will see your students reach new heights in their ability to reason and communicate mathematically. A carefully designed program will breed enthusiasm for mathematics, pride, and quality workmanship. *Writing Math Research Papers* will help you begin your program. It is recommended that you read the Introduction and Chapters 1–9 carefully in order to immerse yourself in the student's frame of reference. Appendix C will help you with the nuts and bolts of setting up different types of program offerings. Undoubtedly, you will make alterations and devise your own system and procedures along the way. We welcome your ideas, recommendations, comments, and questions. You can send them to:

> Key Curriculum Press
> Attn: Editorial
> P. O. Box 2304
> Berkeley, CA 94702

The remainder of Appendix C is divided into the sections shown in the following outline:

I.   When Do Students Write Math Research Papers?
II.   Prerequisites for Math Research
III. Preparation for Math Research
IV. The Teacher as Coach
    A. Consultations
    B. Editing Drafts of the Research Paper
    C. Assessment
       1. Assessing Problem-Solving Skills
       2. Assessing Math Annotation Projects
       3. Assessing Consultations
       4. Assessing Research Papers
       5. Assessing Oral Presentations
V.   Writing Research Papers as an Ancillary Part of a Math Course

If your school does not have a research program in place, you can use this book as a guide to writing a proposal for a new course, or for a program ancillary to existing math courses. If you decide to have students write research papers, read the entire book. Try to make preparations based on the suggestions in Appendix C before the school year begins. Acquire the necessary materials. Choose the articles, problems, and other assignments you will be using initially so the first few weeks of the year are covered. It is helpful to have some articles selected, photocopied, and placed in a folder or binder that students can borrow when they are looking for a topic. The journal articles suggested in Appendix A are a good source of appropriate articles. Ideally, each student should have access to a copy of these articles, which they can sign out just as they do supplemental textbooks. Having the articles ready for the students facilitates their choosing an appropriate topic in an efficient manner. Use the outline above and the corresponding sections that follow as a checklist to guide your preparation. Above all, make sure that you and your students are primed to benefit from the research experience.

When Do Students Write Math Research Papers?

High school students equipped with the proper skills and motivation can write excellent mathematics research papers. Under what circumstances does a student write a math research paper?

1. The student is in a standard mathematics course and does a paper as an enrichment project—perhaps for honors credit.
2. The student is in an independent study program and, in accordance with school policy, chooses to write a math research paper for the major project.
3. The student is enrolled in a mathematics research course elective in which the paper is the central project.
4. The student is presenting a paper as part of a "graduation by exhibition" requirement.
5. The student intends to enter a local or major mathematics contest in the near future or as an upperclass student.

Prerequisites for Math Research

The procedures necessary to foster effective research and writing habits are independent of the class for which the paper is being prepared. The amount of time devoted to reading, writing, and research will affect the quality of the research paper. What are the student prerequisites for success in mathematics research?

1. The student should have successfully completed a standard algebra course typical of most ninth-grade core mathematics programs. Ideally, the course should include a unit on logic. Many of the derivations and proofs students will encounter rely on a knowledge of algebra and deductive reasoning.
2. The student should be a dedicated mathematics student. The student must have good work habits and turn in required work on time.
3. The student should, ideally, be *electing* to write the paper. Self-motivation is crucial, because research requires perseverance. However, like many school requirements, the learning experience can still be valuable if the student is writing the paper solely because it is required.
4. The student should currently be enrolled in a mathematics course.
5. The student should have access to a word processor.
6. The student must attend a one- or two-period orientation session that explains the paper-writing procedures, as delineated in the next section of Appendix C. Scheduling, time lines, and all requirements should be specifically spelled out so students know exactly what demands the research paper will put on their workload. If the paper is being written as an ancillary part of a standard mathematics class, the orientation should take place the very first day of class, before the core curriculum starts. The research, editing, and revision processes require

perseverance and good work habits, so these procedures should be stressed at the orientation session.

Preparation for Math Research

Students meeting the above prerequisites are good candidates for the research paper project. The success of their paper is further ensured if they do some preparatory activities as part of their mathematics research before actually starting on their topic. A mathematics research paper is a major project that should be completed over a period of months, not days. Students involved in such a project should dive in with a "can do" attitude. This self-confidence will come from the student's perception that he or she is ready for the task at hand. Students must be equipped with the tools necessary for basic research. They are apt to be more interested in acquiring these tools if they see their paper as part of a "greater picture" that has purpose, is exciting and challenging, and is perceived as conquerable.

Experience with reading, writing, and problem solving is the necessary foundation. How can students gain this experience in preparation for writing mathematics research papers?

1. The student should become familiar with basic problem-solving strategies, as discussed in Chapter 2: Problem Solving—A Prerequisite for Research.
2. The student should complete a Math Annotation Project, as described in Chapter 4: The Math Annotation Project.
3. The student should complete a Journal Article Reading Assignment, as described in Chapter 7: Reading and Keeping a Research Journal.
4. The student should have access to as many of the sources listed in Appendix A as possible. In particular, the student should read *Writing Math Research Papers* in its entirety (excluding Appendix C) and have it available for reference during the writing project.

Students who meet the prerequisites and complete the preparatory activities are truly equipped to start a research project. Students only meeting the prerequisites can and do write excellent papers, but the mistakes they make along the way would most likely have been avoided if the preparatory activities had been completed.

Students writing a paper without completing the Math Annotation Project or the Journal Article Reading Assignment experience more growing pains, but can still produce high-quality work. Conversely, students can gain much by doing the Math Annotation Project and/or the Journal Article Reading Assignment even if they don't complete a math research paper. The first three preparation activities are more powerful if they are done in a classroom situation. The discussion, questions, mistakes, and comments contributed by a full class of students strengthen the research papers. Below is a sample time

line that coordinates the research paper, the preparation activities, and the reading of
Writing Math Research Papers. Use it to get an idea of when each phase of the research
should begin. Adjust the months proportionately if the research project is to be completed in half a school year. To familiarize yourself with some of the terms used in the
time line, read *Writing Math Research Papers*, Chapters 1–9.

Month	Activities
1	Students read and discuss the Introduction to Writing Math Research Papers and Chapters 1 through 6. They choose a topic and an article and make copies of their article. They can start a bibliography tree and should start a Math Annotation Project.
2, 3, 4	Students read and discuss Chapter 7. They finish the Math Annotation Project and complete the Journal Article Reading Assignment. They begin reading and annotating their article and keeping a research journal. Consultation sessions begin.
5, 6, 7	Students continue research and consultations. They read Chapter 8 and review Chapters 3 and 4 and are ready to begin making an outline for their formal paper. They can start writing, submitting, and revising drafts of the formal paper. They receive teacher feedback on all written work.
8	Students stop any new research and spend time polishing their formal research papers. They submit drafts, make corrections, and discuss the formal paper at consultation sessions.
9	Students read Chapter 9 and prepare an oral presentation and present it to the class.

In some cases, time may not allow for completion of all of the preparation activities
before the research is started. I have coached students who produced excellent papers
written with and without all combinations of the first three activities listed earlier, but
doing them all allows the student to proceed with greater independence. They are valuable tools for doing quality math research. Equipped with them, the student is fine-
tuned for success and, with the teacher as coach, is empowered with the thrill and responsibility of exploration in mathematics.

The Teacher as Coach

Many students have become aware of the fact that they are approaching mathematics in all their classes differently than previous generations approached it. The teacher's role is also changing; perhaps your math department has taken several initiatives in the areas of technology, assessment, problem solving, portfolios, and so on. In this section, we specifically address the teacher's role as coach in math research. As a coach, you want steady progress from your students. To help ensure that progress, you will have periodic consultation sessions with your students, edit each draft of their research papers, and assess the major areas of their project.

Consultations

Students research and write their papers primarily on their own time. (There may be a few instances in which students work on their papers in a classroom situation; these are discussed later in Appendix C.) The teacher acts as coach. You must make sure that students make consistent, gradual progress on a timely basis and that they improve their reading, writing, and research skills. Students need guidance and deadlines to accomplish these goals. For this reason, an essential part of any research program is private consultations between the teacher and the student. Individual students should meet periodically with a mentor (usually the teacher) and discuss the progress on their article. A consultation period is an extra—but mandatory—help period devoted to the paper. (Students might also come for extra help on the paper during your designated extra-help time.)

The consultation session is scheduled by the student and teacher for a mutually agreeable time slot. Before school, after school, lunch period, and free periods are all used to accommodate the consultation. The length and frequency of consultations will depend on both student need and when you can accommodate them in your schedule—but make sure you conduct them as regularly as possible. Some students can touch base with you for five minutes every two weeks to discuss material. Other students may need more time. Consultations should be scheduled so as to ensure that each student is progressing at a reasonable pace. You and your students must commit to this key feature of the research program. The consultation procedure requires the student to engage in research between sessions, preparing material for each consultation. Newly completed work must be brought to each consultation; otherwise, the teacher has the right to schedule a new meeting.

The student discusses his or her new work at each consultation. As a result, over the course of the entire project, the student has worked extensively on the research in many separate sittings. This process is the antithesis of the infamous "paper started the weekend before it is due" syndrome. The strength of the consultation period is the one-on-one communication—there is an onus on the student because there is "nowhere to hide," unlike a normal classroom setting. This raises the students' level of responsibility, their expectations of themselves, and, as a result, achievement. The consultation time is not an extra, a frill, or a dispensable luxury—it is an essential part of the research program.

What happens at each consultation? What is your role? You assign readings from the article(s) the student picked. Make sure you assign appropriate amounts of reading; a paragraph in mathematics can be very involved. Students come to the next consultation with the material read, underlined, and annotated directly on their copy of the article. They should have taken notes that ask questions, explain material from their article, make a conjecture, test a claim, try a proof, or extend the reading passage. To keep track of their current assignment, each student keeps a Consultation Record Sheet, a shorthand journal of their tasks for the week. Appendix C_1 shows a sample completed Consultation Record Sheet followed by a blank one that you can copy and use for your students. Each student needs one sheet per marking period.

The Consultation Record Sheet can be filled in by the student and/or teacher during the consultation period. It clearly delineates what needs to be done before the next consultation. Some items entered on the sheet will be for future reference and not for the very next session. The record sheet sets up a reasonable time line for the students, something they will need to do on their own in college. They will have a method they can adopt and adapt for college use, and you will have a way to keep them progressing in a logical manner with a good deal of student accountability. Consultation Record Sheets from previous marking periods should be saved by the students; they usually contain suggestions for the future that students will need. You should schedule at least seven consultations per student each quarter, depending on your school's scheduling and vacations.

You need a calendar in order to keep track of when students are coming for their consultations. Appendix C_2 shows a sample filled-in Consultation Appointment Calendar followed by a blank one. You can enlarge the blank one and use it—make one copy for each school month. Enter the month on each sheet and staple the sheets together in a folder. Keep the folder in a safe, readily accessible place; it will last the entire year. If either the student or the teacher is absent, the student must sign up in a new time slot in the folder. Students sign up for their next consultation time when they leave their present consultation. There is room on the calendar for both the date and your particular school's cycle number. Some schools have A and B days, and some are on six-day cycles, ten-day cycles, and so on.

Have your own copy of each student's article available for the consultation. At the beginning of each consultation, reacquaint yourself with the student's progress using the consultation sheet. Go over the student's work, ask questions, answer questions, and so on. Discuss the material and assign the next reading. Force students to be explicit in their explanations; this will improve their oral and written communication skills. You can keep a small, inexpensive cassette recorder handy and allow students to tape the consultations. Each student buys a cassette and uses it week-to-week to re-create the entire consultation. Recording the consultations is optional; offer it as a suggestion and let students use their judgment about whether it would be effective for them. The consultation allows you and the student to have a mathematical discussion that will raise the level of the student's content knowledge and communication skills. It is terrific preparation for an oral presentation as well as an excellent forum for witnessing the combination of student empowerment and accountability.

Editing Drafts of the Research Paper

When the students begin formal writing of their individual papers, they will be writing up material they are familiar with. They will devise an outline and follow the writing recommendations in Chapter 3: Writing Mathematics and Chapter 8: Components of Your Research Paper. As they produce pages, they will submit them to you for editing. Papers can be submitted at consultations or at any other time during the school day. Drafts should be submitted a few pages at a time to allow you to read and return them in a realistic length of time. Advise students to have disk and hard-copy backup if they are leaving drafts in your mailbox. As you read the papers, give full-sentence suggestions, make corrections, and ask questions. Acquaint yourself and your students with the editing symbols shown in Chapter 3. These shorthand comments take care of the more mundane corrections. Other comments will require phrases or full sentences and should be made either in the margins or between the double-spaced lines of text. Be sure you check the mathematics as well as the writing. Look over the frequent teacher comments listed in Chapter 3.

You can return edited drafts in class, at the consultations, or during free times during the school day. Students will revise their papers according to your editing, and hand in the revised material along with any new material. This constant "trading" continues until the paper is finished. You can discuss the edits at the consultations. Editing suggestions should appear on the draft, not on the Consultation Record Sheet. Students must save all drafts. Sometimes old drafts are needed, especially if material is lost on the word processor. When the paper is completed, it will not be a "new" reading for you; you will know the material and the writing well because you will have seen it so often. As the coach, you played an important role in the growth of the paper, and you will be assigning a grade for the work done. How will you evaluate the different facets of the research process? Guidelines are given in the next section.

Assessment

Grading short-answer questions is generally considered uncomplicated. Grading geometry proofs and long questions where partial credit is given for work shown requires more thought. Assessing problem solving and research requires a balance between grading specific content and holistic evaluations. The large amount of quality contact time the research teacher has with each student optimizes the teacher's ability to make a reasonable

judgment of the student's progress. We will discuss assessment in five areas:
1. problem-solving skills
2. the Math Annotation Project
3. consultations
4. written papers
5. oral presentations

   Your research program may not include traditional tests as indicators of progress, and rightfully so. Problem-solving and math research skills cannot be timed; their quality is measured by the final product, not by how the work appears after a forty-minute period. Students need to know how they are doing, so make it clear what will be graded. You can base students' quarterly grades on a combination of the five areas listed above, but note that all five may not be part of every marking period. The relative weights of each area are determined by you; make sure they reflect the degree to which the area was stressed during each marking period. Students should be made aware of the specific grading criteria that will be used. We give some suggestions for grading in the five areas; feel free to adapt them to realistically judge the priorities of your particular course.

1. Assessing Problem-Solving Skills

Grading nonroutine problems given in homework, in class, or on tests takes more time than grading short answers or traditional "show all work" problems. Because the grader can't be a mind reader, students must be required to explain each step they take in solving such problems. Grading scratch work that lacks verbal explanation requires the grader to infer what the student was thinking, and this is not a desirable situation in making evaluations. Several books on teaching problem solving include a scoring scheme that makes grading problems more objective. Some are included in Appendix A. Ted Herr and Ken Johnson (1994) adapted a grading strategy suggested by Randall Charles in *Problem Solving Experiences in Mathematics* (1986). We use Herr and Johnson's adaptations to create a list that consists of five 2-point criteria. Each problem then is worth 10 points, and students can just multiply by 10 to get a percent to judge their problem-solving ability. Numbers indicate the number of points that should be awarded.
   A. Understanding the Problem
      0: Student misinterprets the problem completely.
      1: Student misinterprets part of the problem.
      2: Student understands the problem completely.
   B. Choosing a Problem-Solving Plan
      0: Doesn't use any strategy.
      1: Uses an inappropriate strategy.
      2: Chooses a correct strategy that would suffice to solve the problem.

C. Executing the Plan

    0: Doesn't execute any plan, or uses the chosen strategy incorrectly.

    1: Uses the chosen plan, but with some errors.

    2: Uses the plan without any errors.

D. Answering the Question

    0: Doesn't give an answer.

    1: Gives an incorrect answer due to an error in previous steps, or gives correct answer based on erroneous information.

    2: Gives correct answer, based on correct information, in a full sentence.

E. Explanation

    0: Offers no explanation in full sentences, or is too vague.

    1: Gives explanations of all the steps taken with some errors or ambiguity, or is incomplete.

    2: Gives a comprehensive, precise explanation.

For each problem, the grade consists of a total followed by an ordered set of five numbers that correspond respectively to A, B, C, D, and E above. You might want to look at other suggestions for grading problem solving and/or adapt this outline. Show your grading criteria to your students to help them understand their grades better and concentrate on areas that need improvement.

2. Assessing Math Annotation Projects

The Math Annotation Project, featured in Chapter 4, allows students to begin to practice and receive comments on their writing. This practice will make the paper writing a smoother process.

Students earn grades from 1 (lowest) to 5 (highest) for each of the following criteria:

❑ The paper covers all of the notes.

❑ The paper is mathematically correct.

❑ The paper is well organized with respect to sections and paragraphs.

❑ The physical layout of the paper, including diagrams, tables, and word processing, is high-quality.

❑ Diagrams are graduated where necessary.

❑ The material is explained well (proper sentence structure and correct math terms).

❑ Original examples and cited examples not from the class notes are included.

❑ The steps of proofs and derivations are adequately explained.

❑ The examples are appropriate, sufficient, and chosen with purpose.

❑ Good explanations of typical pitfalls are given.

❑ Questions from handouts, tests, and quizzes are included and analyzed.

❑ A list of key terms, correlated with page numbers, is given.

❏ The use of technology (calculator keystroke sequences and computer programs) is explained.

❏ Technology is used with discretion.

❏ The list of writing tips was followed.

❏ The paper was submitted, edited, and revised in a timely fashion.

❏ All recommendations and corrections from edits are incorporated into the paper.

❏ The paper does a clearer job of explanation than the original notes.

❏ The general depth and quality are commensurate with the student's ability; that is, the student worked to his or her potential.

You can convert the raw score into a percent and enter a grade in your gradebook for the annotation project. Add or delete items to make this list accurately reflect your priorities. Students should save their annotation projects along with the drafts to use as reference when they write their research papers. Students can complete more than one annotation project if time permits.

3. Assessing Consultations

Students should be made aware that their performance at consultations will be assessed at the end of the quarter. Individual consultation sessions are not graded. The grade is a holistic grade that reflects effort, depth and quality of questions, depth and quality of extensions, evidence of original thought, work with proofs, article annotations, testing of claims, quality of notes, and punctuality. Students earn grades from 1 (lowest) to 5 (highest) for each of the following criteria:

❏ The student achieves or exceeds Consultation Record Sheet goals.

❏ There is evidence of time and effort spent on the research.

❏ The articles are annotated and underlined.

❏ Notes are taken and material from the articles is reworded.

❏ The student asks good questions.

❏ Each claim is tested via a few examples.

❏ The student makes original conjectures.

❏ The student tests claims made in the reading and/or the original conjectures.

❏ Proofs are attempted and/or completed.

❏ The student makes extensions based on the readings.

❏ The student's work shows evidence of probing and persistence.

You can convert the raw score into a percent and enter a grade in your gradebook for consultations. Add or delete items to make this list accurately reflect your grading criteria. Naturally, you will uncover differences in the level of the students' work. Some students will exceed expectations, while others will fall short in some areas mentioned above. Remind the students that their grade will be based not on occasional exams but on performance as monitored consistently throughout the marking period. In this respect, it is a very realistic indicator.

4. Assessing Research Papers

You will be very attuned to each paper when the final version is handed in because you will have edited it extensively and discussed it in the consultations. Your last reading of each paper is perhaps the easiest, since all the comments you made previously have been incorporated. Keep in mind that you are grading a final paper—try to judge the work itself. It may have required more effort and more pain for some students to get to the same level as other students. If the paper's topic was appropriate for the student, then there is no "penalty" or "reward" for doing many drafts and revisions. The consultation grade can be adjusted to reflect that effort and commitment. For the research paper grade, students earn grades from 1 (lowest) to 5 (highest) based on your evaluation with respect to each of the following statements:

❏ The abstract is clear, succinct, and comprehensive.
❏ The problem is stated clearly.
❏ The paper is mathematically correct
❏ The paper does a clearer job of explaining the material than the original article does.
❏ The student includes and explains original patterns and/or conjectures.
❏ Proofs in the article are explained.
❏ There are original proofs.
❏ The student includes original extensions of some of the paper's ideas.
❏ The paper is well organized.
❏ Diagrams are graduated where necessary.
❏ The physical layout of the paper, including diagrams, tables, and word processing, is high-quality.
❏ All edits and recommended changes are incorporated into the paper. The explanation given for not incorporating any specific recommended change is satisfactory.
❏ The material is explained well.
❏ Appropriate and sufficient examples are given.
❏ The recommendations for future research flow naturally from the completed research.
❏ The general depth and quality of the paper are commensurate with the student's ability; that is, the student works to his or her potential.

The extensive editing over a period of months should "weed out" any problems, and the resulting paper should be high-quality. You might want to weight the last question to count for more than 5 points, for example, from 1 (lowest) to 20 (highest). Compute the raw score, convert it to a percent, and enter a grade.

5. Assessing Oral Presentations

When the papers are completed, you can have your students give oral presentations, as discussed in Chapter 9. The oral presentations can be graded independently of the paper and the consultations. For the oral presentation grade, students earn grades from 1 (lowest) to 5 (highest) based on your evaluation with respect to each of the following statements:

- ❏ The student writes and revises an outline for the presentation.
- ❏ The talk is logically organized.
- ❏ The introduction clearly orients a novice to the essentials of the research.
- ❏ The student creates an appropriate handout.
- ❏ The student speaks in full, mathematically correct sentences.
- ❏ The student allocates time wisely and uses discretion in condensing the paper for the oral presentation.
- ❏ The student finishes on time, at an appropriate juncture.
- ❏ The student can answer questions that are asked during the presentation.
- ❏ The student knows the material and does not rely on "cue cards" or on reading the paper.
- ❏ The board work is appropriate and well organized.
- ❏ The overhead transparencies are effective.
- ❏ The posters are helpful and carefully prepared.
- ❏ The student uses color on the posters and transparencies to convey a mathematical idea.
- ❏ The audience is engaged, participating, and asking questions.
- ❏ The student uses manipulatives and technology effectively and appropriately.

Above all, try not to let the student's performance on the paper and the consultations predetermine the oral presentation grade. You may find that a student with an excellent paper needs work on the oral presentation. Other students may surprise you with excellent visuals, good organization, and a smooth delivery. A copy of a blank grading sheet can be given to students so they will be aware of the grading criteria. Students in schools that require students to graduate by "exhibition" may be able to use their oral presentation of their math research to satisfy this requirement.

Writing Research Papers as an Ancillary Part of a Math Course

Students often become engaged in the research experience as an ancillary part of their core mathematics course. In such cases, the one- or two-period orientation session should take place the first week of school—the first days are recommended. If the entire class is working on research papers, class sessions should be devoted to discussing the chapters in *Writing Math Research Papers* as they are covered. Have students follow a timetable similar to the one in the Introduction. Students who meet the prerequisites can complete the preparation activities, pick a topic, and begin their research. It is advisable to have a selection of articles already photocopied by you from which students can select their topic. This will expedite the choice of topic and get students started quickly. Appendix A lists many excellent articles that can be used to get students started.

Once a topic has been picked, the students need to begin consultation sessions with the teacher. As subsequent chapters are read, they need to be discussed in the classroom or consultation setting. If the entire class is writing research papers, the nine chapters should take nine periods to discuss in class; you can use more periods if time allows. If an entire class chooses to do only the Math Annotation Project and/or the Journal Article Reading Assignment, the suitable chapters should be discussed in class. If the entire class is not writing research papers, the discussion of the chapters read by the research-paper students can be discussed on days when the class is split into groups for performance assessments, projects, and other activities. If the entire class is working on research papers, you can hold consultation days in class, meeting with several students as others work on their research. You might even be able to hold in-class consultations while students are completing other in-class assignments.

If you would like every student to be engaged in math research but have a large class and need to cut down on the potential workload, have students work on their papers in cooperative groups of two. Students will need to meet outside of class to complete this assignment. Each group hands in one paper. You may decide to limit the article choices so several students are working on the same article and can communicate with one another as well as with you. (It may even be possible to find articles that are closely related to the coursework you are doing.) If every student does not write a research paper, have the other students do an independent math project chosen from a list of options. The projects could consist of original ideas and/or ideas taken from publications. The NCTM offers material on mathematics projects, and your library may have books on math projects. In the past, students have written math quiz shows, designed and explained math models, done statistical surveys, and used graphs to create artwork.

You may decide to have all students who don't elect to do a research paper do a Math Annotation Project for a three-week unit studied in math class. Randomly assign each unit to a different student so that, as a class, the entire year is covered. The units can be assigned as they are completed. Ideally, students will take better notes in class, annotating them as they write them down, because they know they have a chance of being selected to do an annotation project. Invariably, each student internalizes the material from his or her individual annotation project and becomes the expert for that unit. These annotation projects can be compiled to form a binder of the entire year's notes that students can use if they are absent, are missing notes, or want extra material to use for studying. This is an effective project that improves the students' writing skills and content knowledge in your class.

If there is time in your schedule, students can give oral presentations of their completed papers in class. You may have a student who would like to present the paper at a math department faculty meeting.

Implementing a One-Semester Course in Math Research

The ideal forum for the math research undertaking is to set up a one-semester course dedicated to mathematics research. Many different scheduling arrangements are possible. You may need to customize some of the following suggestions so they work within the restrictions of your school's schedule.

Scheduling Class Meeting Times

A math research course can be set up as a one-semester, 1/2-credit course. It could meet either every day for one semester or every other day for the entire school year. There are several ways to put in one semester's worth of time over a full school year. If the school has a rotating schedule, the course could meet every "even" day or every "odd" day. Or the research class might meet on days when students have no science lab or no physical education, depending on how your school sets up its master schedule. The longer time spread of the course allows the extra time needed to revise, edit, reformulate conjectures, devise proofs, and so on in between class meetings. If this arrangement is impossible, the course can meet every day for one semester. Although the continuity and concentration levels are high with this format, the students might find it hard to complete independent work on a nightly basis. With this arrangement, several days should be allotted for certain assignments. Discuss with your mathematics department chairperson, principal, and guidance counselors how a math research program can be scheduled in your school. We will discuss a one-semester course titled Problem Solving and Math Research and make suggestions for a one-semester follow-up course called Investigations in Math Research. The first year your school offers Problem Solving and Math Research, the course must be advertised. As a new course, it could easily be overlooked, even if it appears in the school's course catalog. You can create a flyer about the course and have it distributed by math teachers. Students from all grades can sign up.

Problem Solving and Math Research

The curriculum in Problem Solving and Math Research covers the problem-solving strategies discussed in Chapter 2 and the process of writing research papers. The two topics are treated concurrently. This should be explained at the orientation session, which can be held the first one or two class sessions. Students should receive a copy of *Writing Math Research Papers* and be asked to read the Introduction and Chapter 1 for homework that night. Chapter 1 can be discussed in class the following day or two. Let's look at the problem-solving and research paper components of the one-semester course in detail along with assessment for such a course.

❏ **The Problem-Solving Component of Problem Solving and Math Research:** Chapter 2 gets students started on the problem-solving facet of their course. Students should read the chapter for homework, and it can be discussed in class the following day. This introductory course in math research requires the instructor to spend most of the class time on problem-solving strategies. *Problem-Solving Strategies: Crossing the River with Dogs, and Other Mathematical Adventures,* by Herr and Johnson, is a recommended text for the problem-solving component of Problem Solving and Math Research. The first problem-solving sessions of the course feature lessons on George Polya's problem-solving steps. You should use Polya's classic, *How to Solve It,* as a reference; most libraries have this book. *The Art of Problem Posing,* by Brown and Walter, is also a helpful reference. There are many other resources for problem-solving material; some are listed in Appendix A and at the end of Appendix C.

A lesson on each strategy should be presented, with examples, questions, extensions, and homework. You can use a combination of lecture, cooperative learning, and developmental and discovery techniques to build each lesson. The initial presentation of each strategy should include several problems that are solved with that particular strategy. Before students can be expected to choose strategies, they need to become adept at each one. Hopefully, this treatment will allow them to begin to associate types of strategies with types of problems to some degree. After all of the strategies have been covered separately, the homework assignments and class problems can mix up all of the strategies and/or combine several of them.

Nonroutine problems of all types can be given. Don't make a habit of reviewing classwork problems at the end of the period—students shouldn't get into the mindset of "We'll go over it at the end of the period; I'll wait for that." Be creative in finding incentives for them to keep trying. Occasionally you can give hints. Never accept blank paper without evidence of the students' scrutinizing and underlining (see Chapter 2) the problem. Let them list questions they have about the problem if they can't solve it. Students can solve problems cooperatively, independently, in assigned groups, in groups they choose, or sometimes in a setting of their own choice. Some problems may take several sessions to solve; others may take only a few minutes. Each student keeps a notebook just as he or she would in any other class. Students can be required to write formal solutions to problems as they are completed in class. This exercise will help sharpen their writing skills, familiarize them with your editing marks, and give them practice in explaining problem-solving strategies. When the problem-solving component gets under way, look for students who get particularly frustrated with the homework and class problems. These students want to *get the answer*. Becoming used to the frustration and necessary determination that accompany problem solving is not an overnight transition. As the year progresses, the steady diet of problems and their solutions should allow all students to gain confidence. The communication that takes place among students when they are in groups or simply commenting during a class discussion will be very helpful in precipitating success in problem solving.

While the students are solving the problems, you become a roving eavesdropper, asking questions, giving hints and encouragement, and so on. When you review the solutions, you can elicit questions, extensions, and similar problems from the students. Answers are not necessarily given at the end of each period, nor are they necessarily given by the teacher. Your role is to be a discussion stimulator and facilitator. A student can present the solution while you act as a commentator, adding annotations to the student's presentation. At this stage, students' communication skills can be improved by hearing you reword, clarify, and expand upon their explanations. Point out to them that your purpose is to teach them to explain concepts more clearly. As time goes on, ask *them* to reword selected explanations from the problem. Go over homework in a similar fashion. Over a period of months, you should see a dramatic change in the confidence, ability, and zest of your students when they attempt to solve a nonroutine problem. Frustration, determination, and elation will often work together.

Problem-solving homework assignments can be assigned weekly, handed in, and graded. Each homework assignment can include problems you select, plus questions from old Scholastic Aptitude Tests or other standardized tests. The students will then get an early introduction to such tests as part of their problem-solving homework. Familiarizing the students with the standardized tests helps alleviate the anxiety associated with them and allows students to become comfortable with the types of questions they will face and with using their problem-solving skills to answer them. You may decide not to give weekly homework when the students begin to write up their research, because this new skill will occupy much of their time.

Where do you find the problems for students to solve? Appendix A lists several books on problem solving. The monthly calendar of *Mathematics Teacher* is a great source of problems. Your math department's library may already have some problem-solving books. Your school library and public libraries also have books on problem solving. As mentioned in Chapter 2, your classes will create their own problems for a schoolwide publication. Start a bank of original problems by saving each year's problems. Use these problems in subsequent courses. Some newspapers and magazines feature brainteaser columns; clip these and file them. Virtually all of these problem sources offer solutions, answers, or both. Many sites on the World Wide Web computer network offer Problems of the Week. See Swarthmore College's Mathematics Forum (http://www.forum.

swarthmore.edu), for example. As you become more comfortable with the problem-solving sessions in your research class, feel free to assign a problem for which you don't have the answer. Students need to see an experienced mathematician; you, at work—whether you find a solution or not. The problem-solving component of the course can run through the entire year. The problem-solving experience the students receive should help them clear the hurdles they'll face during their research project.

❏ **The Math Research Component of Problem Solving and Math Research:** Students will take breaks from their problem solving as they cover chapters in *Writing Math Research Papers*. Use the time line on page 164 to orchestrate the research activities and the reading of each chapter. As students read each chapter, it must be discussed in class. Students will complete a Math Annotation Project and a Journal Article Reading Assignment, pick a topic, and get started. Familiarize yourself with the section of Appendix A that has sources for paper topics, and photocopy a selection of articles for students to use. Once the individual research has begun, you can, at any juncture, use class sessions to discuss research material. As the students progress in their research and consultations, assign and discuss the chapters in *Writing Math Research Papers*. You can allow students to work on their research in class on selected days. On these days, you can conduct consultations with several students during class. When drafts start being handed in, you might want to spend sessions discussing examples of student work that can help the entire class. Toward the end of the course, students can prepare oral presentations and actually present them in class.

❏ **Assessment:** Grades are based on problem-solving exercises, homework, paper writing, class participation, and consultation quality. Assessment suggestions are given in Appendix C. Formal tests and/or take-home tests can be given at the teacher's discretion, and they can include questions about problem-solving theory and/or research techniques. The time limitations of in-class tests hinder your ability to give nonroutine problems. You may give problem-solving tests to students in cooperative groups, even over a period of days. Such tests could involve a multistep application problem that serves as a performance assessment.

Problem Solving and Math Research can be an enriching experience. The articles, the problems, the consultations, the writing, the oral presentations, and your effort and encouragement can transform a high school student into a budding researcher. If students in Problem Solving and Math Research are also in a core mathematics course, this tandem provides them with a solid, balanced mathematical experience.

A One-Semester Follow-Up Course in Math Research

If enough students would like to continue their math research, you may decide to offer a one-semester follow-up course to Problem Solving and Math Research. This follow-up course, called Investigations in Math Research, could be repeated for credit as long as the students pick a new topic or further investigate a topic explored in a previous paper. The routine for the individual paper is the same as in Problem Solving and Math Research—students pick a topic for their individual paper and attend consultations. The in-class agenda, however, is different.

Investigations in Math Research will have the students working on research topics in-class every day. This daily in-class routine differs from the problem-solving emphasis of Problem Solving and Math Research. You can offer students in this follow-up course several options for their in-class routine:

1. the Cooperative-Paper Model
2. the Buddy-System Model
3. the Solo-Concentration Model

We discuss each of these options in detail in the following sections. You may decide to have all three options at work in the class simultaneously, or you may pick one option and have the entire class follow that model. Feel free to experiment with your own adjustments of these suggestions. If you need to conduct consultations during the class period, you can easily work them into the in-class research, since all students are always engaged in their research.

The Cooperative-Paper Model

Students who opt for (or are assigned) the Cooperative Paper Model work on two separate research papers, one on their own and one cooperatively. The in-class time is devoted to mathematics research and the production of the cooperative paper. You can either assign groups or let the students choose their own groups. Base this decision on your knowledge of the students' work in Problem Solving and Math Research. Groups of two or three students work best. The group must choose an article that is different from the articles the group members are working on individually and in consultation. The cooperative paper will have no required consultation period. It is completed entirely, including the research and the word-processing, in the class period.

The teacher is a roving consultant during the class period. As you move around the room helping students during the research phase, you will become attuned to which students are actively involved, questioning, probing, and so on. Because all students are not working on one short problem (as in a problem-solving lesson), you will need to spend extended time with certain groups at certain times. Allow students to interrupt your interaction with a group to call you over for answers to quick questions they have. Keep a copy of each article being researched for your reference. As you enter a group, ask them to explain their most recent progress to you. This will orient you and require them to practice their oral communication. Ask questions of different members of the group. Ask to see their notes, journals, scratch work, and article annotations. Remind students that their daily performance in these groups is a determinant of their grade and is always being assessed. Comment on their work. Tell them what is going well and what needs work. Make sure they stay on task; friends can easily slip into a conversation that slows down progress on the research. Require each student to keep a journal and notes. In some cases, as students work together on a proof, one person might transcribe it while the others pitch in ideas. Their journals must correctly account for the location of such material; it is advisable that all students have all the notes in case something is lost. This type of class period, with its student-as-worker, teacher-as-coach format, will keep you very busy.

Each student in a group must have a copy of the group's article. For the first half of the course, the students read through the article, verifying claims, constructing proofs, and writing questions, extensions, and conjectures. Each student must also keep a journal of the progress made on each article. If a student is absent or there is a long vacation period, the journal is invaluable in keeping the group up-to-date. Appendix C_3 shows a sample completed journal form followed by a blank one. Take time to look at these sample pages. By the middle of the course, students are ready to write their papers. At this point, the class can meet in the computer lab. The new twist is that they must figure out a way to combine their strengths to create the best paper possible. Logistically, they also must figure out writing assignments for the group. The editing process is exactly like the editing process for the individual papers. You can address questions and edit pages as they are written or as students ask questions. Let the students watch you edit some pages and explain your comments to them as you write them. This will help students when they proofread their own writing; they will try to imagine your reaction if they are familiar with your expectations. The teacher remains a roving coach and can help students by answering questions about the material even as it is being written on the computer screen. The completed paper is handed in for a grade. This grade, along with the student's individual paper and research, is the basis for the student's final course grade. You can holistically assess each student and incorporate this assessment into their final grade as well. Interaction with the students on a regular basis can build a relationship of respect, camaraderie, and high regard for quality scholarship.

Often, the students are really "getting into" their research when they need to start writing it up, and stopping new research to start writing the formal paper may seem counterproductive. If you are confident that quality research is being done and students are keeping excellent notes and journals, consider accepting legible handwritten work as verification of their work and let them keep researching. They are still honing their writing skills on their individual paper. If the strength of cooperative research and its resulting communication is proving to be effective, let it continue. Students completing Investigations in Math Research participate in two research papers during the course: their individual paper and the cooperative paper. Combining this experience with their

Problem Solving and Math Research experience, students have not only learned research skills, they have gained research experience. Students may repeat Investigations in Math Research for credit if they choose different topics or extend their current topic. If an intact-group elects to repeat Investigations in Math Research, they can continue their cooperative paper.

The Buddy-System Model

The class time in the Buddy-System Model is spent on another cooperative venture, one different from the Cooperative-Paper Model. Students work in teams of two and meet in these groups during each class. They will assume the roles of commentator, author, researcher, and editor. We will explain the procedure using the example of Samantha and Matt.

The first day, both students work on Samantha's paper. Matt has a copy of Samantha's article, and Samantha has explained her progress to Matt. Matt asks questions about material he doesn't understand or about conjectures he has. These discussions progress according to the individual progress Samantha has made on her paper. As Samantha begins to write, Matt can help her edit her written work before it is submitted to the teacher. (The teacher still edits any submitted drafts.) Matt is familiar with the paper and the article, but he does not work on it during any of his own time. He is engaged with Samantha's topic during Samantha's class time. Since Samantha does independent work at home and has one consultation per week, Matt "falls behind." Samantha must bring Matt up-to-date. She cannot proceed with her research until she can explain it to Matt. (She probably shouldn't be proceeding until this stage anyway.) This procedure will help Samantha determine what she doesn't know, communicate more clearly, and internalize the material more quickly. (How often do teachers remark that they didn't understand some topic fully until they were forced to *teach* it?) For example, it might help her practice fielding questions that will be posed by readers and judges at math contests if that's a goal she has for the project. Matt will be questioning her on a steady basis either to have his questions answered or to force Samantha to formulate a pristine presentation. The only drawback is that if Samantha starts to make tremendous progress between class meetings, she will have much to explain to Matt, and the class sessions may start to lag behind her current progress. It is still an effective procedure.

On the days when the tandem is not focusing on Samantha's research, the roles reverse, and Matt's paper becomes the topic of their discussion. Samantha plays Matt's role as described above. At the end of the year, students give an oral presentation to the class. Samantha gives her presentation on an aspect of Matt's work that can be taught comfortably in the allotted time. Samantha makes an audience-participation handout (see Chapter 9). Roles reverse for Matt's presentation. Note that the students are not writing a second paper—they concentrate on their individual topic, with their buddy contributing feedback.

Contrast the cooperative research done under the cooperative-paper option with that done under the buddy-system option. In the cooperative-paper format, students are on an equal footing; they are coproducers of the cooperatively written paper. In the buddy-system format, the relationship changes. The person who provides the feedback for the researcher has to be able to question, critique, challenge, and yet work with and encourage the researcher. These roles reverse on alternate class meetings, so the students get a chance to empathize—they'll be on both sides of the fence. This requires a sophisticated, professional relationship, and for that reason it may not be advisable to allow students to work with friends. On the other hand, when people who work together associate with each other outside of school, their topic often surfaces in impromptu conversation, and they may be spending time outside of class discussing mathematics. Since each relationship is different, you must be the judge of how the student pairs can most effectively be set up. Notice that these two options can actually be used in one class, in the same room, at the same time. You will play the same role for the students in both formats. The third suggestion, the Solo-Concentration Model, does not involve cooperative work.

The Solo-Concentration Model

The Solo-Concentration Model has students spending their in-class time on their individual research. This model is best reserved for students who are preparing to enter a contest and want to concentrate exclusively on their own research. It lacks the dependency on communication that the other models feature. You may decide to restrict this option to students who are continuing work on a paper previously written. Students can work within this model in the same Investigations in Math Research class in which other students are following the other two models. The solo-concentration students attend consultations and give an oral presentation to the class.

A variation of this model is the independent study option, which does not require the student to attend any classes, just consultations. You may decide to reserve this avenue for students who are able to make progress on their own. If an especially talented student is progressing with original material, it is possible that the knowledge of the teacher-consultant has been exhausted. Such a student may need a mentor. Other teachers, the department chairperson, community people, businesspeople, and local college professors can serve as mentors. Many colleges support outreach programs that may be able to assist you in this endeavor. Partnerships between high schools and colleges have been developed. Students can perform work at the college, or the high school may take on a student teacher from the college as part of the mentorship agreement. If a student reaches such a level, you will need to creatively problem-solve in order to fulfill the student's need for high-level feedback. Ideally, the talents of these budding mathematicians should be harvested.

The suggestions for one-semester courses in Problem Solving and Math Research and

Investigations in Math Research are based on experiences with these two research courses at North Shore High School over the years 1991–1996. There is much latitude, and you may create your own custom version of a research course by combining aspects of the programs suggested. If you have financial concerns about beginning math research as an ancillary program or a separate course, the next section may give you some ideas.

Funding Your Research Program in Its Infancy

As you can see, math research can be pursued in a variety of ways. Formal programs require more financial support because they count as class assignments in a teacher's schedule. All programs need books, audiovisual materials, contest entry fees, poster materials, access to technology, mentors, and so on. If your school decides to offer a formal math research program as part of teachers' schedules, you'll need some time to develop the program. Any form of a research program will require summer curriculum work to gather the program materials. If financial support becomes an issue, you should consider options for raising funds. Become aware of local, state level, and national grants you may be eligible to apply for. Establishing a research course could make you eligible for a grant. The NCTM's Mathematics Education Trust offers several scholarships for teachers. For information on scholarships currently being offered, write to:

> National Council of Teachers of Mathematics
> Department E
> 1906 Association Drive
> Reston, VA 22091

You can also write to the federal and state governments and request information about funding opportunities. Many agencies and addresses are given in *Guidebook to Excellence—A Directory of Federal Resources for Mathematics and Science Education Improvement*. This book can be acquired from:

> Eisenhower National Clearinghouse
> The Ohio State University
> 1929 Kenny Road
> Columbus, OH 43210-1079

You can purchase a copy of *Teacher Fellowships, Scholarships and Awards: Finding and Winning Funds for Professional Development* by writing to :

> Capitol Publications
> P.O. Box 1453
> Alexandria, VA 22313-2053

Your local or state mathematics teacher organization may also offer funding. Read their newsletters and journals regularly to find announcements of such opportunities.

Some expenses can be paid for by the students. They can pay to photocopy the articles they plan to research and to make extra copies of their final paper. You may decide to have students pay for art supplies required for posters. Some students may elect to subscribe to *Mathematics Teacher*, *Quantum*, *Mathematics and Informatics Quarterly*, or another math publication described in this book. Special subscription rates are usually available to students.

Research classes sometimes take day or overnight trips to science museums, math fairs, math teachers' conferences, lectures, and special events. Some colleges allow high school students to "shadow" college students to get a taste of college classes. In some cases, students can stay overnight in the dormitories. Contact the colleges you are interested in for details. Students are also welcome, at a reduced registration fee, at NCTM Regional and National meetings. Registered students can attend sessions, and you can require them to report to the class about what they learned. A Regional or National Meeting may at some point be held in your area; look in the NCTM journals for advance information. Find out if your state or county math teachers' meeting allows students to attend. Student volunteers are usually needed to help out at meetings; your students may be interested in this. If you are requiring students to pay for transportation, meals, admission fees, or hotel costs, be sensitive to their ability to afford the trip you plan. Consider fund-raising activities similar to the fund-raisers held by other school clubs.

Beginning a Math Research Library

Appendix A lists many publications that are appropriate for a math research library. You should try to accumulate as many items as you can by adding to your library each year. A math resource library can be set up as part of the school library or as part of the math research classroom or departmental office. In addition to the books and periodicals, a catalog of completed student papers could be set up as a resource for students wishing to *extend* a given paper rather than write their own paper on the same topic. Start your own library of photocopied articles that are good sources for research topics. Search for articles in old journals in your school or local college library. Make lists of readings from resource books that contain potential research topics. Cross-reference the readings and articles that are related. As material accumulates over the years, start a card catalog of research topic ideas for students to use as they search for a topic.

Resource Books

Many of the books in the following list are specifically written for teachers, but can also serve as valuable resources for students. Some are general interest books about mathematics, some are about problem solving, and some offer help with research writing.

Brown, S., and Walter, M. *The Art of Problem Posing*. Hillsdale, N.J.: Lawrence Erlbaum Associates, 1990.

Chazan, D., and Houde, R. *How to Use Conjecturing and Microcomputers to Teach Geometry*. Reston, Va.: NCTM, 1989.

Countryman, J. *Writing to Learn Mathematics*. Portsmouth, N.H.: Heinemann, 1992.

Davis, P., and Hersh, R. *The Mathematical Experience*. Boston, Mass.: Houghton Mifflin, 1981.

Flegg, G. *Numbers. Their History and Meaning*. New York: Schocken Books, 1983.

Gardner, M. *The Incredible Dr. Matrix. The World's Greatest Numerologist*. New York: Charles Scribner's Sons, 1976.

Guillen, M. *Bridges to Infinity*. Los Angeles, Calif.: Tarcher, 1983.

Lehoczky, S., and Rusczyk, R. *The Art of Problem Solving—Volume 1: The Basics*. Stanford, Calif.: Greater Testing Concepts, 1993.

Lehoczky, S. and Rusczyk, R. *The Art of Problem Solving—Volume 2: And Beyond*. Stanford, Calif.: Greater Testing Concepts, 1994.

Maletsky, E. *Teaching with Student Math Notes 1*. Reston, Va.: NCTM, 1987.

Maletsky, E. *Teaching with Student Math Notes 2*. Reston, Va.: NCTM, 1993.

Mathematical Sciences Education Board and National Research Council. *Reshaping School Mathematics. A Philosophy and Framework for Curriculum*. Washington, D.C.: National Academy Press, 1990.

National Council of Teachers of Mathematics. *Curriculum and Evaluation Standards for School Mathematics*. Reston, Va.: NCTM, 1989.

National Council of Teachers of Mathematics. *Professional Standards for Teaching Mathematics*. Reston, Va.: NCTM, 1991.

National Research Council. *Everybody Counts. A Report to the Nation on the Future of Mathematics Education*. Washington, D.C.: National Academy Press, 1989.

Paulos, J. *Innumeracy. Mathematical Illiteracy and its Consequences*. New York: Hill and Wang, 1988.

Polya, G. *How to Solve It*. Princeton, N.J.: Princeton University Press, 1973.

Polya, G. *Mathematical Discovery: On Understanding, Learning and Teaching Problem Solving*. New York: John Wiley and Sons, 1981.

Sachs, L. *Projects to Enrich School Mathematics, Level 3*. Reston, Va.: NCTM, 1988.

Schaaf, W. *The High School Mathematics Library*. Reston, Va.: NCTM, 1987.

Steen, L. *On the Shoulders of Giants. New Approaches to Numeracy*. Washington, D.C.: National Academy Press, 1990.

Whimbey, A., and Lochhead, J. *Beyond Problem Solving and Comprehension*. Hillsdale, N.J.: Lawrence Erlbaum Associates, 1984.

Whimbey, A., and Lochhead, J. *Problem Solving and Comprehension*. Hillsdale, N.J.: Lawrence Erlbaum Associates, 1982.

Catalogs of Resource Books

Some catalogs feature books that are excellent resources for research papers. You can send for the catalog and acquire books as your budget allows. In some cases, you may be able to preview the book before purchase. In preparation for your research library, send for the following catalogs:

Creative Publications
5040 West 111th Street
Oak Lawn, IL 60453

Undergraduate Mathematics Applications Project (UMAP)
High School Mathematics Applications Project (HiMAP)
COMAP
60 Lowell Street
Arlington, MA 02174-1295

Dover Publications
31 East 2nd Street
Mineola, NY 11501

W. H. Freeman
41 Madison Avenue
New York, NY 10010

Janson Publications
P.O. Box 860
Dedham, MA 02027-0860

Key Curriculum Press
P.O. Box 2304
Berkeley, CA 94702

Mathematical Association of America
1529 Eighteenth Street, NW
Washington, DC 20036

National Council of Teachers of Mathematics
1906 Association Drive
Reston, VA 22091

Dale Seymour Publications
PO Box 10888
Palo Alto, CA 94303-0879

Springer-Verlag
175 Fifth Avenue
New York, NY 10010

Wadsworth School Group
10 Davis Drive
Belmont, CA 94002-3098

J. Weston Walch
321 Valley Street
P.O. Box 658
Portland, ME 04104-0658

Starting a Math Contest in Your Locality

Appendix C_4 lists general information about well-known math and science competitions. For various reasons, all students may not be able to enter these contests. Perhaps the deadlines conflict with other student commitments. Perhaps the school cannot support a research program, or the hiring of a mentor. Maybe the students have other priorities and cannot devote the time and energy needed to enter a competition at the level of the Westinghouse Talent Search, for example. In any case, your research students should all have a chance to give their oral presentations in a formal forum.

Several districts and counties have math competitions in place already; schools have math teams that are part of interscholastic math leagues that focus on problem solving. The network of these schools is an excellent starting point for a local math research contest. A committee of interested teachers could meet, set up rules, and create an annual competition. New York has an excellent model for this. In the Al Kalfus Long

Island Math Fair, students participate in two rounds: a preliminary round in March/April and a final round in April/May. All students who submit papers are separated according to grade level. The fair includes students in grades 7–12. The students participate in the preliminary round by submitting their papers and giving a fifteen-minute oral presentation to a room with one to three judges and five to eight student presenters. The judges volunteer; they are teachers, businesspeople, professors, and community members. The event is held at a local college. After the students in the room present their papers and answer judges' questions, several students are selected from each room to participate in the second round. At the second round, the preliminary-round procedures are repeated with the selected students. Winners of gold, silver, and bronze medals are chosen from each room and announced at an awards assembly. Specific questions about the fair can be sent to the address listed in Appendix C_4. If you are starting a math fair in your community, consider using videotapes of the oral presentations as submissions with the papers if it is impossible to arrange the "live" event.

In addition to, or in lieu of, the competition, your school or district can hold a Math Research Night at the high school. Teachers, students, parents, and community businesspeople should be invited, and refreshments can be served. Artwork from the oral presentations can be on display before and after the presentations. Students can give abbreviated versions of their oral presentations, with a special effort to convey their work in an effective way to people who generally will not be mathematicians. The presentations can be given by one student at a time to a seated audience, or each student can have a table in the gymnasium to present their materials. Attendees then can mill around the gym from table to table, asking students about their work. Students are available to speak about their work for the entire evening. The math department could issue awards to students based on criteria established before the event. Money from admission tickets can be used to help defray the costs of the research program, such as books, field trips, materials for students' posters, and so on. The event can be held annually in the spring semester. The school community is used to watching theater events, sports events, and musical events performed by students. Certainly, a student performance displaying the academic power, knowledge, and enthusiasm acquired through a math research project epitomizes the academic mission of the school and deserves to be at centerstage.

Problem Solving & Math Research

Consultation Period Record Sheet Marking Period **3** Name **DANIÈLE**

Session	Date / Time	Notes
1	Jan 5 / Pd. 5	Read p. 464 first column. Test claim #4 on p. 464. Write up explanation of Thm. 2 on p. 463.
2	Jan 12 / 7:40	Rewrite proof on p. 465 if 'n' was not prime. Read # at bottom of p. 466. Give an example.
3	Jan 18 / 7:40	Fix; comments on draft. Rewrite p. 4 of draft to include your explanation of your extension to Thm. 3, p. 467.
4	Jan 26 / 3 pm	Fix most recent draft. Extend idea on p. 469 to include perfect squares. Try Fibonacci Numbers. Write up orig. proof from p. 468.
5	Feb 2 / 7:20	Make table to show pattern in Thm.6. Use spreadsheet. Make Conjecture if x is odd only. Try to prove conjecture.
6	Feb 8 / 7:20	See if Thm. 8 on p. 471 holds for trinomials with irrational roots. Fix last draft. Finish reading article 1.
7	Feb 16 / Pd. 5	Start reading article 2; read p. 172-173. Apply findings from article 1 to last # on p. 172. Test all claims.

Problem Solving & Math Research

Consultation Period Record Sheet   **Marking Period** _____   **Name** _____

Session	Date	Time	Notes
1			
2			
3			
4			
5			
6			
7			

RESEARCH CONSULTATION CALENDAR   14 STUDENTS   MONTH: JANUARY

Monday	Tuesday	Wednesday	Thursday	Friday
1 (A) Tim 7am Sarah Pd 5 Katie 7:20	**2** (B) Nicole 7:20 Jeff Pd 2 Sandy 7:40	**3** (A) Sharon Pd 4 Matt 7:20 Simran 7:40	**4** (B) Anthony 3pm Robin Pd 6	**5** (A) Danny Pd 5 Jocelyn 7:40 Blake 7:20
8 (A) Tim 7am Sarah Pd 5 Katie 7:40	**9** (A) Jeff Pd 2 Sharon 3pm Nicole 7:20	**10** (B) Sandy Pd 7 Matt Pd 2 Simran 7:40	**11** (A) Anthony 3pm Robin Pd 6	**12** (B) Danny 7:40 Jocelyn Pd 5 Blake 3pm
15 (B) Tim 7am Sarah Pd 5 Katie 7:40	**16** (B) Sandy Pd 7 Nicole 7:20 Sharon 7:40	**17** (A) Jeff Pd 2 Matt 7:20 Robin 7:40	**18** (B) Anthony 3pm Danny 7:40 Simran Pd 4	**19** (A) Jocelyn 7:20 Blake 3pm
22 (B) Tim 7:40 Sharon 7:20 Sarah 3pm	**23** (A) Jeff Pd 2 Katie 7:40 Nicole 7:20	**24** (B) Sandy Pd 7 Matt 7:40 Simran 3pm	**25** (A) Anthony 7:40 Robin Pd 6 Jocelyn 7:20	**26** (B) Danny 3pm Blake 7:40
29 (A) Tim 7:20 Sharon 3pm Katie 7:40	**30** (B) Jeff Pd 2 Matt 7:40 Nicole 7:20	**31** (A) Sandy Pd 7 Sarah Pd 5 Simran 7:40		

Research Consultation Calendar

Month:

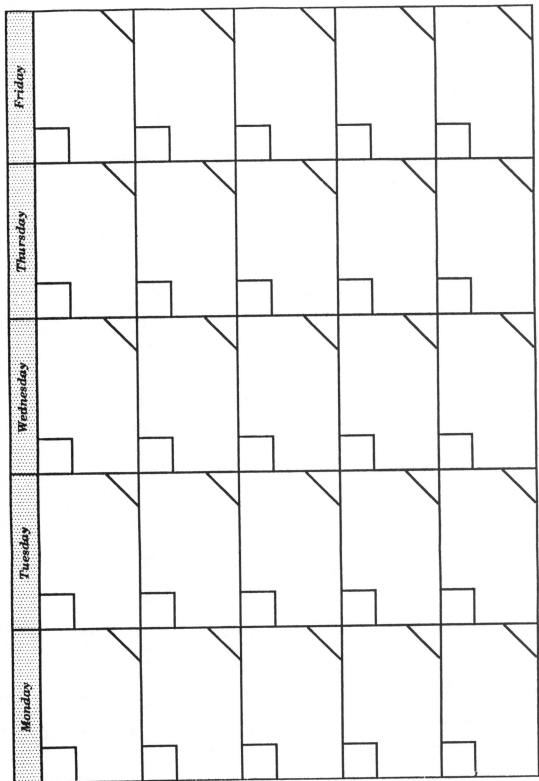

PROGRESS JOURNAL FOR COOPERATIVE PAPER

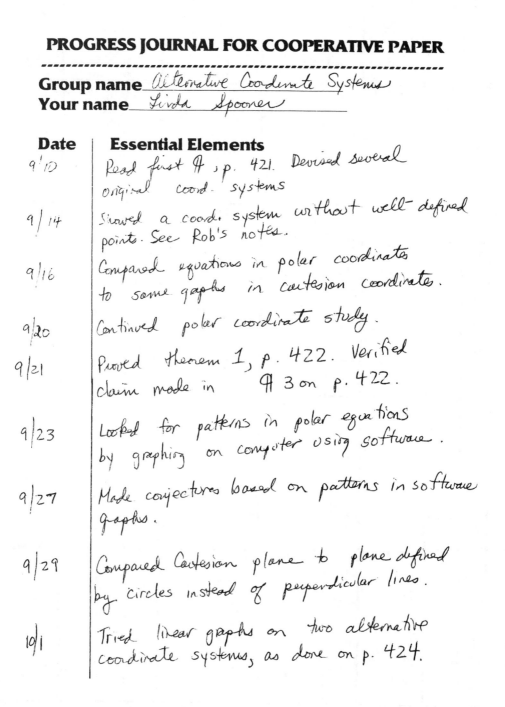

Group name *Alternative Coordinate Systems*
Your name *Linda Spooner*

Date	Essential Elements
9/10	Read first ¶, p. 421. Devised several original coord. systems
9/14	Showed a coord. system without well-defined points. See Rob's notes.
9/16	Compared equations in polar coordinates to some graphs in cartesian coordinates.
9/20	Continued polar coordinate study.
9/21	Proved theorem 1, p. 422. Verified claim made in ¶ 3 on p. 422.
9/23	Looked for patterns in polar equations by graphing on computer using software.
9/27	Made conjectures based on patterns in software graphs.
9/29	Compared Cartesian plane to plane defined by circles instead of perpendicular lines.
10/1	Tried linear graphs on two alternative coordinate systems, as done on p. 424.

PROGRESS JOURNAL FOR COOPERATIVE PAPER

Group name_____

Your name_____

Date	**Essential Elements**

Appendix C₄—Contest Addresses

The Al Kalfus Long Island Math Fair
c/o Joseph Quartararo
273 Cambon Avenue
St. James, NY 11780

American Regional Math League
c/o Bronx High School of Science Math Department
75 West 205th Street
Bronx, NY 10468

American Regional Math League Power Contest
Columbia Heights High School
1400 49th Avenue NE
Columbia Heights, MN 55421

Continental Mathematics League
P.O. Box 5477
Hauppauge, NY 11788-0121

DuPont Science Essay Awards Program
General Learning Corporation
60 Revere Drive
Northbrook, IL 60062-1563

Junior Engineering Technical Society (JETS)
1420 King Street, Suite 405
Alexandria, VA 22314-2794

The Mandlebrot Competition
Greater Testing Concepts
P.O. Box A-D
Stanford, CA 94309

Math League Press
P.O. Box 1090
Manhasset, NY 11030

Mathcounts (seventh- and eighth-graders only)
1420 King Street
Alexandria, VA 22314-2794

NASA Space Science Student Involvement Program (SSIP)
NASA Headquarters
Education Division
Mail Stop FEO
Washington, DC 20546

National Mathematics League
P.O. Box 9459
Coral Springs, FL 33075

Duracell Scholarship Competition
National Science Teachers Association
1840 Wilson Boulevard
Arlington, VA 22201-3000

Toshiba Exploravision Contest
NSTA
1840 Wilson Boulevard
Arlington, VA 22201-3000

Odyssey of the Mind
P.O. Box 547
Glassboro, NJ 08028
USA Mathematical Talent Search
Department of Mathematics
Rose-Hulman Institute of Technology
Terre Haute, IN 47803

The Gelfand Outreach Program in Mathematics
c/o Harriet Schweitzer
Center for Math, Science and Computer Education
SERC Building, Room 239
Rutgers University—Busch Campus—Box 1179
Piscataway, NJ 08855-1179

International Science and Engineering Fair
Science Service
1719 N Street, NW
Washington, DC 20036

Tandy Technology Scholars
P.O. Box 32897
Fort Worth, TX 76129

American High School Mathematics Examination
Dr. Walter E. Mientka, Executive Director
Department of Mathematics and Statistics
University of Nebraska-Lincoln
Lincoln, NE 68588-0658

The USA Mathematical Olympiad
Dr. Walter E. Mientka, Executive Director
Department of Mathematics and Statistics
University of Nebraska-Lincoln
Lincoln, NE 68588-0658

The Westinghouse Science Talent Search
Science Service
1719 N Street, NW
Washington, DC 20036

National Science and Humanities Symposium
Academy of Applied Sciences
98 Washington Street
Concord, NH 03301

References

Boulger, W. Pythagoras meets Fibonacci. *Mathematics Teacher*, vol. 82, no. 4, pp. 277–282.

Brown, S. From the golden ratio and Fibonacci to pedagogy and problem solving. *Mathematics Teacher*, vol. 69, no. 2, pp. 180–188.

Brown, S., and Walter, M. *The Art of Problem Posing*. Hillsdale, N.J.: Lawrence Erlbaum Associates, 1990.

Campbell, D., and Stanley, J. *Experimental and Quasi-Experimental Designs for Research*. Boston: Houghton Mifflin, 1966.

Charles, R. *Problem Solving Experiences in Mathematics*. Reading, Mass.: Addison-Wesley, 1986.

Consortium for Mathematics and Its Applications. *High School Lessons in Mathematical Applications*. Lexington, Mass.: COMAP, 1993.

Countryman, J. *Writing to Learn Mathematics*. Portsmouth, N.H.: Heinemann, 1992.

Dynkin, E., and Uspenskii, V. *Problems in the Theory of Numbers*. Lexington, Mass.: D. C. Heath, 1963.

Ganis, S. Fibonacci numbers. In *Historical Topics for the Mathematics Classroom*. Reston, Va.: NCTM, 1969, pp.77–79.

Gannon, G., and Converse, C. Extending a Fibonacci number trick. *Mathematics Teacher*, vol. 80, no. 9, pp. 744–747.

Gibb, G., Karnes, H., and Wren, F. The education of teachers of mathematics. In *A History of Mathematics Education in the United States*. Reston, Va.: NCTM, 1970.

Hansen, D. On the radii of inscribed and escribed circles. *Mathematics Teacher*, vol. 72, no. 6, pp. 462–464.

Herr, T., and Johnson, K. *Problem Solving Strategies—Crossing the River With Dogs, and Other Mathematical Adventures*. Berkeley, Calif.: Key Curriculum Press, 1994.

Huntley, H. *The Divine Proportion*. New York: Dover, 1970.

Kelly, L. A generalization of the Fibonacci formulae. *Mathematics Teacher*, vol. 75, no. 8, pp. 664–665.

Kimmins, D. The probability that a quadratic equation has real roots: An exercise in problem solving. *Mathematics Teacher*, vol. 84, no. 3, pp. 222–227.

Kline, M. *Mathematics in Western Culture*. London: George Allen & Unwin, 1954.

Lyubomir, L. What is the use of the last digit? *Mathematics and Informatics Quarterly*, vol. 1, no. 1, pp. 15–17.

Milanov, P. On the Malfatti problem for equilateral triangles. *Mathematics and Informatics Quarterly*, vol. 2, no. 2, pp.47–53.

National Council of Teachers of Mathematics. *Curriculum and Evaluation Standards for School Mathematics*. Reston, Va: NCTM, 1989.

National Council of Teachers of Mathematics. *Professional Standards for Teaching Mathematics*. Reston, Va.: NCTM, 1991.

Olson, M. Odd factors and consecutive sums: An interesting relationship. *Mathematics Teacher*, vol. 84, no. 1, pp. 50–53.

Paulos, J. *Innumeracy. Mathematical Illiteracy and Its Consequences*. New York: Hill and Wang, 1988.

Polya, G. *How to Solve It*. Princeton, N.J.: Princeton University Press, 1973.

Polya, G. *Mathematical Discovery: On Understanding, Learning and Teaching Problem Solving*. New York: John Wiley and Sons, 1981.

Raphael, L. The shoemaker's knife. *Mathematics Teacher*, vol. 66, no. 4, pp. 319–323.

Russell, B. *A History of Western Philosophy*. New York: Simon & Schuster, 1957.

Schielack, V. The Fibonacci sequence and the golden ratio. *Mathematics Teacher*, vol. 80, no. 5, pp. 357–358.

Sgroi, R. Communicating about spatial relationships. *Arithmetic Teacher*, vol. 37, no. 6, pp. 21–23.

Stark, H. *An Introduction to Number Theory*. Chicago, Ill.: Markham Publishing Co., 1970.

Steen, L. *On the Shoulders of Giants. New Approaches to Numeracy*. Washington, D.C.: National Academy Press, 1990.

Stewart, I. *The Problems of Mathematics*. Oxford: Oxford University Press, 1987.

Tirman, A. Pythagorean triples. *Mathematics Teacher*, vol. 79, no. 8, pp. 652–655.

Vakarelova, V. On the Bobillier theorem. *Mathematics and Informatics Quarterly*, vol. 2, no. 1, pp. 34–35.

Vorobyov, N. *The Fibonacci Numbers*. Lexington, Mass.: D. C. Heath, 1963.

Weinberg, S., and Goldberg, K. *Statistics for the Behavioral Sciences*. Cambridge: Cambridge University Press, 1990.